THIS BORROWED EARTH

THIS BORROWED EARTH

LESSONS FROM THE FIFTEEN WORST ENVIRONMENTAL DISASTERS AROUND THE WORLD

Robert Emmet Hernan

Foreword by Bill McKibben
Preface by Graham Nash

 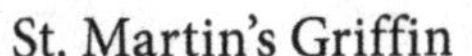 St. Martin's Griffin  New York

www.stmartins.com

Designed by Newgen Imaging Systems (P) Ltd., Chennai, India

Library of Congress Cataloging-in-Publication Data

Hernan, Robert Emmet.
This borrowed earth : lessons from the fifteen worst environmental disasters around the world / Robert Emmet Hernan.
p. cm.
Includes bibliographical references and index.
ISBN 978-0-230-61983-8 (trade paperback)
ISBN 978-0-230-10527-0 (e-book)
1. Environmental disasters—Case studies. I. Title.
GE146.H37 2009
363.7—dc22 2009019880

Our books may be purchased in bulk for promotional, educational, or business use. Please contact your local bookseller or the Macmillan Corporate and Premium Sales Department at (800) 221-7945, extension 5442, or by e-mail at MacmillanSpecialMarkets@macmillan.com.

First published by Palgrave Macmillan, a division of St. Martin's Press LLC

First St. Martin's Griffin Edition: February 2010

D 10 9 8

CONTENTS

Acknowledgments vii

Foreword by Bill McKibben ix

Preface by Graham Nash xi

Introduction 1

Minamata, Japan, 1950s 9

London, England, 1952 31

Windscale, England, 1957 39

Seveso, Italy, 1976 45

Love Canal, New York, 1978 61

Three Mile Island, Pennsylvania, 1979 81

Times Beach, Missouri, 1982 91

Bhopal, India, 1984 101

Chernobyl, Ukraine, 1986 109

Rhine River, Switzerland, 1986 129

Prince William Sound, Alaska, 1989 137

Oil Spills and Fires of Kuwait, 1991 155

Dassen and Robben Islands, South Africa, 2000 163

Brazilian Rainforest 171

Global Climate Change 179

List of Some Environmental Organizations 189

Notes 193

Sources 197

Index 231

ACKNOWLEDGMENTS

My thanks and deep appreciation to:

Jeanne Silverthorne, my wife and best friend, for just about everything;

Bill McKibben for providing the clearest and most compassionate voice in support of efforts to save this planet from those who don't care;

Graham Nash for his efforts throughout his career to support environmental protection, for his support for this project from its earliest incarnation, and for his forceful preface;

My friends and supporters Bill Kulik, Jim Morris, and Carl Garvey, for generously sharing their thoughts and comments after reading the manuscript;

Luba Ostashevsky, my editor at Palgrave Macmillan, for her support and thoughtful comments throughout the editorial process, and her assistant, Laura Lancaster, for her help in editing and finalizing the photo rights;

Linda Langton, my agent, of Langtons International Agency, for her enthusiasm and effort to place the book in difficult times;

Elizabeth Durst for her editing that tightened and strengthened an earlier version of the text;

The important photographers and organizations that generously donated permission to use their photographs for free or at reduced cost to support the project, including Robert Del Tredici, Dino Fracchia, Jim West, Greenpeace, International Fund for Animal Welfare, and Shisei Kuwabara;

The members of environmental groups who taught me the importance of citizen activism and government enforcement, including Rick Brown and Tom O'Leary and the Lower Providence Concerned Citizens; Flare Deegan, Marge Tombler, and Jeannie Losagio of Save Our Lehigh Valley Environment; and Peter Bulka, Lois Gibbs, and Debbie Cerrillo at Love Canal;

The people of Minamata, Japan, who have endured with strength and faith the suffering inflicted on the community for the past fifty years, and those who serve to bear witness to that suffering, including Hanaga-san, Michiko Ishimure, Aileen Smith, Eiko Sugimoto, Jun Ui, Satoshi Kamizawa, and the staff at Minimata Disease Center Soshisha.

The author purchased fifteen tons of carbon credits to offset the estimated carbon emissions resulting from the publication of this book. The calculations were derived from "Going Carbon Neutral: A Guide for Publishers (U.S. Edition)," www.newsociety.com. The carbon credits for a reforestation project were purchased through carbonfund.org.

FOREWORD

Are these events really receding into distant memory? It seems impossible to those of us who lived through them, but I think it's true. Some college students I quizzed recently had only a dim recollection of hearing about Chernobyl, and Bhopal elicited a complete blank. I didn't bother asking about London and its choking fogs.

And so Robert Hernan provides a deep service by reminding us how out of kilter things can go. In an age where we're once again ideologically committed to "loosening the reins" on private enterprise, it's sobering to remember what has happened in the past. In an age when new technologies are barely tested before they're put into widespread use—genetically engineered crops, for instance—it's even more sobering to contemplate a seemingly iron-clad rule: every new machine or system seems to fail catastrophically at least once.

In the years to come, the line that Hernan draws in his introduction between natural and environmental disasters will blur—if global warming raises the sea level and then amps up the hurricane, is the wave that inundates Miami "environmental?" It's not an act of God, that's for sure. Still, in some way it will be an act of collective folly—not the individual and corporate greed that so often seems to stand behind these tragic tales. And it won't respond in quite the same way to the individual heroism that Hernan documents so movingly in these pages.

What will remain the same, however, is human vulnerability. W. Eugene Smith's pictures of the victims of Minamata Disease capture that vulnerability at its deepest, and so do the stories of the firemen at Chernobyl and the many others chronicled herein. That vulnerability endangers us, of course—but it is closely related to the love, the shared concern, that might save us yet. I was speaking not long ago on a panel with Lois Gibbs, the hero of Love Canal, and I found myself thinking how magnificent it was that she had rallied not only to her own cause and that of her neighbors, but also to the victims of a thousand other tragedies. In the end, Gibbs's witness—and the witness of many like her—is more telling than the greed and recklessness of the powerful that created the need for her work (not a lot more telling, but just a hair.)

May this book give heart and courage to many more such great souls, for there are assuredly many more such fights to come.

Bill McKibben

PREFACE

I am deeply saddened by the devastating stories and the tragic images presented in *This Borrowed Earth*. This remarkable book is a great place to begin finding out more about what greedy, incompetent, short-sighted people are doing to me, to you, to our children, and to our fragile planet. It documents some of the "crimes against humanity" and reminds us that we only have one planet to live on.

Like a frog in a pan of water we have not noticed the effects of all this pollution until now, when we feel the water beginning to boil... Aren't we supposed to be custodians of this planet? Don't we owe it to our children to leave them a world that they can live in safely and enjoy?

We must all try our very best to bring this madness to the attention of like-minded people and to the public at large. *This Borrowed Earth* will serve as a wake-up call to all concerned citizens of this earth. If we don't take heed, this unrelenting fouling of our collective home will, without question, come back to haunt us now and in the future.

GRAHAM NASH of *Crosby, Stills, Nash & Young*
Los Angeles, California
May 2009

THIS BORROWED EARTH

INTRODUCTION

Several generations have been born since many of these environmental disasters occurred, and those generations' knowledge of the events is limited. Even for those who were around for some of these disasters, the details of what happened are distant memories. While some might recall the immediate impact of the disasters, there has been little exposition of their long-term consequences, including the health effects that continue to plague those who were exposed to the toxic releases.

If we forget how and why these disasters happened and what horrible consequences emerged from them, we will not avert future disasters. Looming on the horizon is the threat of global climate change caused by the emission of carbon dioxide and other greenhouse gases. Some continue to deny this threat, as others denied the possibility that oil tankers, chemical and nuclear plants, and landfills could leak. In recent years, however, a clear consensus has developed within the scientific and international policy community that global warming as a result of human action is real, and it is upon us. Like Chernobyl, this threat extends across the entire planet, with potential consequences that range from costly to devastating.

To reduce our dependence on fossil fuels (oil, coal, gas) that contribute substantially to global climate change, many stress the critical need to develop renewable sources of energy (solar, wind, biogas). Others encourage a return to nuclear power. In considering nuclear power as an option, it is critical that we remember and learn from the events at Chernobyl, Windscale, and Three Mile Island, the three environmental disasters that occurred at nuclear power plants.

While my main interest is simply to relate the compelling stories of what happened during these environmental disasters, I also feel it is important to highlight certain lessons that have emerged from them. Environmental disasters differ from natural disasters. Natural disasters arise from natural forces, such as floods, tornadoes, hurricanes, tsunamis, and volcanic eruptions. These natural disasters can, and often do, wreak havoc over a wide area and cause deaths in the hundreds, thousands, and even hundreds of thousands. They tend to happen suddenly, have a severe impact, and then recede quickly, in hours or days.

Environmental disasters, in contrast, arise from manmade forces. They pollute the environment—air, water, or land—for months, decades, or, in the case of some radiation elements, thousands of years. Like natural disasters, some environmental disasters can happen quickly, as at Chernobyl and at Seveso, Italy, where an explosion released toxic chemicals into surrounding communities, forcing the evacuation of hundreds of families. Others can develop slowly over time, as at Love Canal in New York, where toxic chemicals seeped into the ground over decades and surfaced in a neighborhood, first threatening the residents, then finally driving them away. In Minamata, Japan, the disposal and release of poisonous mercury into the sea spanned several decades and inflicted unspeakable suffering on entire communities.

Environmental disasters sometimes kill people outright, as at Bhopal, India, where thousands died almost instantly after an explosion at a chemical plant spread a cloud of toxic fumes over the city. The more distinguishing characteristic of environmental disasters, however, is that they often produce their worst health effects only months or years afterward. It was many years after the 1986 nuclear disaster at Chernobyl, for example, that children began to suffer from cancer of the thyroid as a result of their exposure to radioactive materials. The consequences of environmental disasters are in many ways more insidious than those of natural disasters.

Because these environmental disasters are manmade, there is almost always someone to blame, usually the polluter or an ineffective government agency. Not surprisingly, we react differently to suffering that is inflicted by other people rather than by natural forces, especially when the suffering results from a company's callous failure to protect the environment in which we all live. If we are injured or lose property as a result of a flood or hurricane, we don't get angry at the water or wind. But if a company's chemicals contaminate our water and threaten our health and that of our children, we get damned angry. And we want the threat stopped, by the company or by the responsible government agencies.

Environmental disasters can also be distinguished from industrial accidents. While industrial accidents are also manmade, they tend to develop quickly and be short lived, and they pose little lasting danger to the environment. An accident at a munitions factory is a good example: Someone makes a mistake or material is defective, and an explosion occurs, killing workers and destroying property. The damage is over in a short time span. Accidents kill quickly; environmental disasters invade peoples' lives for years.

These distinctions are not hard and fast, and they overlap at times. Bhopal, Seveso, and Chernobyl were industrial accidents, but their wide-reaching effects, their impact on the environment, and their long-lasting health effects set them apart from other industrial accidents.

While each environmental disaster unfolds differently, there are certain patterns of human action that set in motion the events that cause such turmoil. Sometimes someone acts carelessly, or negligently, as in the Exxon Valdez disaster: after the captain of an oil tanker drank too much alcohol and abandoned his duties, 11 million gallons of spilled oil damaged an entire ecosystem. Sometimes the conduct of a polluter rises to criminal recklessness, as in the case of the mercury poisoning at Minamata. In that case, the chemical company knew for years that its wastes were causing horrible deformities among the people of the fishing communities, and yet its managers continued to dump mercury wastes into the nearby fishing grounds. At times someone acts deliberately, as when retreating Iraqi soldiers set oil wells on fire in Kuwait during the 1991 Gulf War.

Often the careless or reckless act that triggers a disaster is not simply a mistake or an accident, but comes about as a consequence of inadequate training, outmoded equipment, or insufficient staffing. Such inadequacies are often the result of a company trying to reduce costs. At Times Beach, Missouri, a commercial firm saved money by hiring an unqualified waste hauler to dispose of their toxic material. At the nuclear facility at Windscale, England, in 1957, a rush to meet production deadlines led to the release of radioactive materials onto the countryside.

When something starts to go wrong in a company's operation, those directly involved often deny that anything really disastrous is happening, thus exacerbating the problem. At Three Mile Island, in Pennsylvania, operators of a nuclear power plant ignored indications on their instruments that something was wrong. Failing to appreciate the imminent danger, the plant continued to operate and released radioactive materials onto an unsuspecting countryside.

As the author Barry Commoner forcefully argues, many environmental disasters are the inevitable outcome of the technological developments that produce modern conveniences.[1] When toxic chemicals emerge from manufacturing processes, they have to go somewhere; when oil is shipped long distances in single-hull tankers, a breach of that hull releases large quantities of oil; when energy is produced by controlling nuclear fusion, human error can cause the loss of that control. We live with the by-products of what we manufacture.

As clear as the causes of these environmental disasters may appear to be, the consequences often remain uncertain. Some are visible, such as the fires in the Kuwaiti oil wells, or the dense fog that blanketed London in December 1952. But with the by-products of modern technology, the harmful substances are often invisible, such as the dioxin in the soils of Times Beach or the toxic chemicals dumped into Love Canal and the Rhine River. In many situations, the full reach of toxic chemicals cannot be determined. At Seveso and Love Canal, authorities could not accurately identify how far into the community the dangerous material had spread.

When a flood strikes, or a hurricane blows through, the cause of our suffering is immediately evident. But when we learn that our water or air has been poisoned and we cannot taste or see that poison, we often do not know the extent of our risk or what further danger awaits us. Such uncertainty is intensified by the fact that disaster-caused illnesses often do not manifest themselves until years or decades after an accident. Sometimes, when communities are given inadequate information, they react in ways that exacerbate the suffering of the victims of these disasters. In Minamata, before the communities learned of the mercury poisoning as the source of their troubles, they assumed that the mysterious disease was communicable and therefore alienated those who suffered from it.

Environmental disasters are deeply disruptive to communities in numerous other ways. They often require the relocation of entire communities from their homes, sometimes permanently. Psychological stress, whether from being uprooted from a community or from grappling with the uncertainty of disease, can be nearly as debilitating as the physical harm. The emotional toll remains one of the hidden costs of environmental disasters.

The turmoil that accompanies environmental disasters erupts in a fairly predictable pattern. The initial consequences are often immediate and severe; they are followed by a lull; then finally the devastating consequences emerge. When the famous London fog descended, it covered everything and everyone with a thick cloak of black soot. It was gone within a few days, to the great relief of everyone. Yet, several months later, officials determined that more than 4,000 people had died as a result of the severely polluted air. At Three Mile Island, a nuclear accident prompted an emergency declaration, and fears of a possible meltdown led to an evacuation. The threat of a meltdown was later deemed a false alarm, but later still authorities realized that the initial emergency was far more serious than had first been imagined. Confusion often reigns in such situations, and events are not always what they seem.

As uncertainty sets in, some will invariably minimize the dangers, again, in part, to reduce costs. Polluters have a vested interest in attempting to reassure the public and regulatory agencies that a situation is not as bad as it may first appear. Regulatory agencies often lack the financial resources to determine the extent of a risk, or to do what is necessary to protect the public. Paying for the health effects of such disasters can dwarf even the astronomical costs of cleaning up environmental disasters. Though more than $480 billion was spent to clean up Chernobyl, Belarus now spends close to 20 percent of its gross domestic product every year on costs related to the disaster.

Disasters often occur because a particular industry or a single company dominates a local economy, and as a consequence, the governmental authorities fail to provide oversight or enforcement to correct operational deficiencies. Chemical companies controlled local economies in Minamata, Seveso, Love Canal, Bhopal, and in Basel, Switzerland, near the Rhine River. Private and government-owned nuclear facilities dominated the economies in Windscale, Three Mile Island, and Chernobyl. In each case, government agencies could not or would not provide adequate oversight or enforcement. As a result, companies failed to install adequate safety equipment, failed to maintain safety systems, and ignored warning signs. People living near the facilities suffered the consequences.

Without adequate environmental laws and regulations, companies generally choose the least costly way to operate. At Love Canal and Times Beach, the companies disposed of highly toxic waste through the least expensive method. The economic goals of a particular government or political ambitions can also lead to cost-cutting and increased risk of disaster. The British government's ambition to become a nuclear power blinded it to certain risks at the Windscale nuclear plant. Soviet economic goals led to careless operations at Chernobyl. Successive governments in Brazil sanctioned the exploitation of the rainforest, while ignoring the environmental consequences of their policies.

Ordinary citizens also participate in the destruction of our environment. A complacent reliance on dirty but cheap fossil fuel contributed to the London fog of 1952 and continues to produce global warming that threatens the entire world community. While these stories of environmental disasters raise haunting images, it is important to recognize that the disasters also occasion heroic behavior from the most ordinary citizens. In the early 1980s when I was in private law practice, I became involved in representing citizen groups who were trying to persuade government agencies to confront the dreadful conditions of landfills near where they lived. The deep conviction, courage, and persistence of these people in their fight to protect their environment and their lives taught me the critical need for environmental enforcement and for active citizen participation in that effort.

I was drawn to the ways in which these environmental disasters unfolded, but also to the ways in which people responded to them. As is evident in their stories, ordinary people refused to allow the disasters to destroy their lives; they refused to remain quiet while others dismissed their concerns. When the polluters tried to deny the effects of their actions, and government officials tried to minimize the dangers, these ordinary citizens gathered the facts on their own. They rallied other citizens to join their effort. They created networks with other groups. They sought help wherever they could—from experts, the wider community, government workers—and they learned whom to trust and whom to suspect. They persisted.

When African penguins nesting on Dassen and Robben Islands off the coast of South Africa were threatened by an oil spill, local environmentalists issued a call for help. The response from around the world was immediate and generous, and the ensuing penguin rescue was dramatic. At Chernobyl, firemen, policemen, doctors, and nurses rushed to the scene of that nuclear inferno to help, and some paid with their lives. In Missouri, two citizens secretly followed a waste hauler for more than a year, documenting where he sprayed and dumped waste, including the town of Times Beach.

In addition to individual action, wider communities respond to environmental disasters with demands for greater protection. Almost every disaster has been followed by calls for government regulation and enforcement. The London fog of 1952 led to the British Clean Air Act of 1956, which finally addressed the long-term fouling of London's air. Love Canal was instrumental in the passage of the 1980 U.S. Superfund law, which provided funds to clean up abandoned toxic waste sites and aggressive enforcement powers to make sure polluters pay for cleanups. The careless disposal of dioxin materials from Seveso led to the European Union's Seveso Directive, which now regulates the transnational shipment of toxic wastes. It often takes an environmental disaster to overcome the normal stagnation and vested interests that block regulatory reform.

The stories that follow demonstrate the critical role of active citizen participation in the protection of our environment. But a cautionary note must be sounded. Voices still dismiss efforts to protect and improve our environment with claims that environmental protections are unnecessary and much too costly. Some argue that concerns over polluted air and water are exaggerated and that we cannot cut back on the use of fossil fuel until it is absolutely certain that global warming is upon us. These stories and the lessons they teach us help to dispel such dismissive responses. The environmental disasters portrayed here are real. They put large populations at substantial risk. The full consequences of each of these disasters could have been avoided. Preventive measures would have cost a fraction of what was spent on cleaning up the accidents and on protecting people from further risks.

The refusal to devote sufficient resources to environmental protection efforts is not just a policy choice; it is not just a means of rewarding interests that are tied to polluting technologies or industries. More dangerously, it is the result of an inability or failure to envision the consequences of our actions. As these stories demonstrate, those responsible for controlling lethal substances are often the ones who fail to imagine that their actions could put people at risk, or they ignore the enormity of the risk. The chemical company in Minamata refused to admit or accept that its waste could be causing the devastating disease inflicted on nearby fishing villages. In other cases, bound to routine, nuclear operators could not believe the readings on their instruments when they signaled critical dangers.

These stories reveal the causes of environmental disasters, how they affect people in deeply disturbing ways, and how critical it is to protect our environment. If we want to avoid further environmental disasters, we need to be vigilant: watch, learn the facts, organize and network, get help wherever we can, and persist.[2] If we are not careful, and if we do not take into our own hands the responsibility for preventing environmental harm, further environmental disasters are inevitable. At this moment, the specter of global warming threatens to destroy us like an avenging angel. We are all vulnerable on this borrowed earth, and we must protect the environment that nurtures us and on which all life depends.[3]

ROBERT EMMET HERNAN

MINAMATA, JAPAN
1950s

Minamata is a fishing town beautifully situated on a bay in the foothills of the mountains on Kyushu, the southernmost of Japan's four main islands. In 1908, the Chisso electrochemical company established a plant there. Labor and land were cheap, and water from the mountains supplied plenty of hydropower.

The company began by using calcium carbide to make acetylene, a fuel for lamps, and then developed facilities for making nitrogen fertilizer and other products. The fertilizer was important for Japanese farming, and it turned into a major export product when the First World War disrupted supplies from Europe. After the war, Chisso developed organic chemical compounds to produce a variety of materials, including acetaldehyde, which employed mercury as a catalyst. Acetaldehyde, first made in 1932, was used in plastics, pharmaceuticals, photographic chemicals, and perfumes. The company prospered as a result of the economic reconstruction following World War II and the Korean War. By the 1950s, the company reemerged as a dominant force in Minamata. Increased production of acetaldehyde and other organic chemical products resulted in a concomitant increase in wastewater, which Chisso continued to dump into Minamata Bay.

Already in the 1920s fishermen had complained about the pollution of their fishing grounds in Minamata Bay, but Chisso was a major source of jobs and revenue and was able to make small payments to the fishermen in return for the right to continue polluting. By the 1950s the waste disposal practices began to catch up with the company as the pollution's impact began to be felt by inhabitants of Minamata and neighboring villages.

Fishing was always a critical resource in Minamata. What the fishermen did not sell, they and their families and neighbors ate. In the early 1950s, mullet, lobster, and shad began to disappear from the once-fertile fishing grounds. Dead fish were found floating on the sea; birds dropped dead from the sky. The local fishermen had to borrow money to eat and to buy nylon nets in order to capture what few fish were left. Nets were often lifted out of the sea bearing only a heavy sludge from Chisso's wastewater. The cats in the village started to dance crazily, bash themselves against walls, jump into the sea, and drown.

In 1954, Dr. Hajime Hosokawa, director of the hospital at the Chisso plant, began to see patients with impaired nervous systems. Mostly fishermen and their families, they had difficulty walking and talking and suffered wild mood swings. Their bodies were racked with convulsions. Most disturbing, newborns were exhibiting symptoms, which indicated the presence of a congenital form of the disease. Local health officials conducted a survey of physicians in the area and found that scores of patients had presented similar symptoms and that many of them had died. Especially affected were the fishing communities south of Minamata, where several members of the same family were often afflicted.

Eiko Sugimoto was born in Modõ, a small fishing community just south of Minamata. Her father was the boss of the local net fishery, and as the only child, though a girl, she was expected to carry on the business. One day in 1958, she returned home from a fishing trip and found her mother confused and unable to light her cigarette; the floor was covered with matches. Sugimoto's father took her mother to the hospital. Since this strange illness appeared to affect members of the same family, neighbors feared the disease might be contagious. When Sugimoto walked down to the shore to care for their boat, she was stoned by her friends and neighbors. Covered with cuts and bruises, she tried to find comfort and safety in a neighboring house. Instead, her neighbors threw excrement on her. Shopkeepers refused to touch the diseased; they made them pass their money in special baskets or leave it on the floor so it could be picked up with chopsticks. When Sugimoto and her father also felt sick, they hid it. Treated like lepers in their own communities, the victims felt deeply ashamed. Recriminations destroyed once-close fishing communities.

As the disease spread through the mid-1950s, suspicions fell on Chisso's wastewater since it was widely known to have ruined the fishing grounds

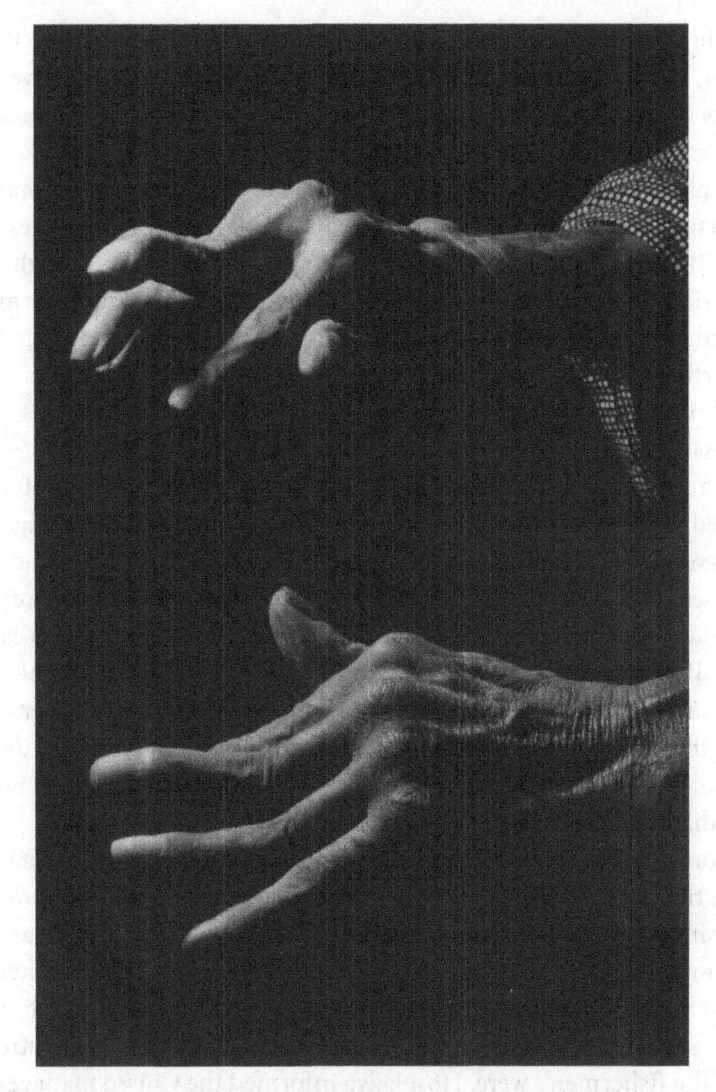

The gnarled hands give testimony to the effects of the mercury poisoning.
Credit: Shisei Kuwabara, courtesy of the artist

in the area. But no one knew what was in the wastewater, and Chisso was not providing any information. Researchers struggled with studies of a host of pollutants found in the bay and were not able to isolate any particular toxic material that would cause such a disease.

In late 1958 a British neurologist who had visited Minamata suggested in *The Lancet* that the disease's symptoms were similar to those produced by organic mercury poisoning. Organic, or methyl, mercury concentrates in the brain and attacks the central nervous system, killing brain cells and turning the brain into a sponge, full of holes. Since it destroys nerve cells, there is no cure for severe cases. The poison can kill a victim in weeks, or slowly eat away at the body for years.

Within a year, a pathologist, Dr. Tadao Takeuchi at Kumamoto University, confirmed the findings, and a special governmental research committee also found that organic mercury was the cause, although it did not attribute the mercury's origins to Chisso's operations. The government disbanded the committee as soon as the report was issued and transferred any further research to a group under the control of several trade ministers who were sympathetic to the company.

Chisso executives tried to deflect attention away from its wastewater by advancing its own theory that the disease was caused by ammunition dumped in the sea at the end of World War II. A researcher at Kumamoto University, Dr. K. Irukayama, discovered, however, that inorganic mercury used as a catalyst in the production of acetaldehyde in the factory was converted into organic mercury. He concluded that the illness was caused by the discharge from Chisso's wastewater, which contained organic mercury. Chisso disputed the charge and claimed that its wastewater could not be the cause since it used only harmless inorganic mercury in its production. The company did not share samples of its wastewater, so no one could disprove the claim.

No one, that is, except Chisso's own Dr. Hosokawa. The doctor had been conducting a series of experiments on cats by feeding them food sprinkled with various chemicals from Chisso's processes. When he fed wastewater from the process that produced acetaldehyde to a cat, it exhibited the same symptoms as those afflicted with Minamata Disease. An autopsy of the cat and lab results confirmed that the cat's cerebellum was destroyed, just as the fishermen's were. Hosokawa informed the Chisso management of the disturbing discovery. The officials ordered Hosokawa to stop his experiments, and the company destroyed all the cats. No replication of the experiment was allowed.

Not only did Chisso deny that its production wastewater was responsible for the disease and suppress Dr. Hosokawa's evidence, the company also steadily increased its production of acetaldehyde and the mercury-laden wastewater. Production in 1950 was 450 tons per month; by 1956 it was 1,325 tons per month; and by 1958 it had increased to 1,500 tons per month. When the water in the sea near the point where the wastewater was discharged south of the plant became heavily polluted, the wastewater was diverted into the mouth of the Minamata River, north of the plant. Dr. Hosokawa warned Chisso against this diversion, but the company ignored him. Within a year, the disease emerged in fishing villages north of Minamata. Yet Chisso continued to increase the manufacture of acetaldehyde with its mercury byproduct, reaching 4,000 tons per month in 1960.

As the disease spread, it became clear that it was related to the consumption of fish that had been contaminated with some toxic substance. Fishing catches had decreased by 90 percent since the outbreak of the disease, and what few fish were left in the area were seldom sold. At first the public simply stopped buying it. Later, the local government barred the sale of fish from the area, which only aggravated the fishermen's financial plight. Fishermen began to hold demonstrations to protest Chisso's destruction of their fishing grounds.

The patients who suffered from Minamata Disease also began to organize. They camped out in front of Chisso's plant and conducted peaceful sit-in demonstrations with the help of a tent donated by Chisso union workers. The patients demanded financial support from Chisso to pay for medical and living expenses. Chisso dominated the economy of Minamata, contributing over half of the city's tax revenue and over one-third of the jobs, and most of the local city officials were former Chisso employees. Because of this, most Minamata citizens were unsympathetic and even hostile to the patients. Through the intervention of the local government, Chisso agreed to a two-part settlement. In November 1959 Chisso agreed to pay the fishing cooperative of 7,000 families ¥35 million ($98,000)[1] as a lump sum compensation, but deducted ¥10 million ($28,000) for damage to its property during one of the demonstrations. Each family ended up with an equivalent sum of about ten dollars. Chisso also provided ¥65 million (about $180,000) for restoration of the fishing grounds, but this was in the form of a loan to the fishermen's cooperative. Then in December 1959, Chisso agreed to also settle with the patients by offering a take-it-or-leave-it deal: ¥30,000 per year ($83) support for each child, ¥100,000 per year ($276) support for each adult, and a lump sum of ¥300,000 ($833) for each dead person, of which there were about 30.

Chisso offered the money only as a *mimaikin,* or condolence, rather than as a *hosokin,* or compensation. In Japan, the condolence is offered as a gift to those less fortunate, as a form of charity, in contrast to compensation, which reflects an admission of responsibility for the harm. Moreover, it was traditional for those receiving a condolence to be grateful and to never again ask for more. Through this deal, the company was also spared the embarrassment of having to ceremonially apologize to the victims.

As part of the settlement, Chisso received a release from the patients to the effect that if proof ever emerged in the future that identified Chisso's wastewater as the cause of the illness, the patients would be precluded from receiving more money from the company. The patients were unaware at the time that Chisso already had the proof, from Dr. Hosokawa and his cat experiments, that the wastewater was indeed the cause of their suffering. For seven more years Chisso discharged over 500 tons per year of mercury-contaminated waste into the sea.

Since no one except Chisso's managers knew of the continuing disposal of the toxic material, most people of Minamata believed that the problem had been resolved after the settlements in November and December 1959. Although more people began to show symptoms of the disease, the fishermen's union pressured its members not to report any new incidence of the disease in order to prevent further damage to the reputation of Minamata's fishing resources. Families discouraged members from identifying themselves as patients since it brought disgrace to the entire family. Culturally, the misfortunes of individuals and families were inextricably intertwined.

Not everyone, however, was willing to deny the existence of the disease. One woman, Michiko Ishimure, came to be the voice of the victims of Minamata Disease through her chronicle of the suffering of the victims, *Paradise in the Sea of Sorrow: Our Minamata Disease*. Ishimure's family was from the Amakusa Islands, across the Shiranui Sea from Minamata, and, like many others, her family had left the islands in search of work. While her grandfather and father were skilled stonemasons, the grandfather's business ran into difficulties, and Ishimure was raised with few physical comforts.

Ishimure's family lived near a cemetery and crematory, an isolation hospital, and a brothel. Ishimure visited the crematory and watched the smoke rise as the dead—epidemic victims, strangers, and the poor—were burned. She spent hours with the prostitutes—girls from poor fishing villages—sitting in their laps as they had their hair done by the local hairdresser. On occasion, she dressed like the prostitutes and paraded along the road. Her grandmother was blind and mad (unrelated to the mercury poisoning), and she went into fits: groping, clutching, and crying out with inhuman noises. She often disappeared from the house. Ishimure would go out looking for her so often that people referred to Ishimure as her grandmother's shadow and to the grandmother as Ishimure's little play doll.

On one occasion the Emperor of Japan was scheduled to visit the Chisso plant in Minamata. All vagrant and deranged people were to be relocated to a small island for the length of the visit. When a policeman came to inform Ishimure's father that his mother would have to be removed, he refused, saying that he would kill himself rather than suffer such a disgrace. The grandmother was allowed to stay, and Ishimure vowed to follow her father's fearlessness in standing up for those, like her grandmother, who needed protection.

Ishimure married a laborer who became a primary school teacher. A son, Michio, was born, and Ishimure supplemented their meager income by bartering on the black market, trading what little fish could be caught for food supplies. Ishimure also began to write, and she published her first verse in 1953. It was about her blind, mad grandmother, and about herself:

> If I go mad, like Grandma, I too
> May be kicked out of the house
> Bodily

The dark poems did not sit well with many, but through her writing Ishimure met Gan Tanigawa, a young revolutionary poet and an activist in the Japanese Communist Party. Tanigawa's fierce, uncompromising demands on literature and society toughened Ishimure. And she needed it, for she was about to discover the afflicted of Minamata.

Tanigawa's literary circle also included Satoru Akazaki, who worked for the city of Minamata. Akazaki obtained, without permission, a copy of what became known as "The Red Book," a journal kept by doctors at a secret ward at Minamata Hospital, where the victims of the strange disease were quarantined. The details of the assaults on the victims from this disease formed an excruciating story for Ishimure.

Ishimure's exposure soon became more immediate. Her son contracted tuberculosis and was admitted to Minamata Hospital. The TB ward was right next to the ward housing the Minamata victims, and she heard terrifying howls from the ward and saw fingernail scratches along the walls.

Ishimure saw patients who were unconscious, and others lying motionless, staring into space with wide-open eyes. Kama Tsurumatsu,[2] a 56-year-old fisherman, who seemed to be little more than a skeleton, frequently fell out of his bed. It was his falling out of his boat while fishing that had led his family to hospitalize him. Yet his eyes still pierced Ishimure, perhaps aided by the sunken cheeks in which they were set. The pain and sadness tore at Ishimure.

Ishimure also met Yuki Sakagami, Patient No. 37, also called Yukijo. Yukijo started fishing when she was three years old, and she was a gifted fisherwoman. Even when fish began to disappear from the sea around Minamata, she could find fish for herself and her new husband, Mohei. Mohei bought a new boat when they got married, and they spent their time together fishing in the sea that Yukijo considered her garden. Mohei was silent and warm; Yukijo was outgoing, often playing, singing, and dancing with children in the neighborhood.

After they had been married for only two years, in the mid-1950s, when Yukijo was forty-one, the symptoms arrived. At first she dropped laundry that she was carrying on her back, without knowing that she had dropped it. Then her hands and legs started to go numb, and she began to stumble. Before long she could speak only in fragments, struggling with each word. Yukijo could no longer help Mohei with the fishing. She thought that perhaps these difficulties were due to fallout from American and Chinese nuclear bomb tests, or the result of early menopause. The latter issue was eliminated when Yukijo became pregnant.

While in the hospital, Yukijo had an abortion because the doctors concluded that her life was threatened. After the procedure was done, fish was served for dinner. Yukijo thought the fish was her lost child. She spoke to it, touched it fondly, and ate it, thinking she was eating her dead child. When a group of dignitaries visited the ward, Yukijo went into a convulsion during which she suddenly shouted, "Long Live the Emperor!" followed by a rendition of the national anthem. The visitors fled. But the behavior served her well with other audiences, for she took to visiting the nearby TB ward, dancing like the cats used to, and singing. She earned cigarettes this way and amused the patients. Yukijo spoke of herself as a "rickety, half-insane, drooling weirdo patient," but at least she was only half insane. The half that remained sane accounted for the loneliness and isolation that she felt and expressed to Ishimure.

Ishimure began to organize on behalf of the patients. She visited the patients and wrote down their testimonies and the horrors she witnessed. But she paid a price for her involvement: Her family received threatening letters, and her parents put pressure on her to spend more time on wifely duties, caring for the household rather than for strangers. Eventually her

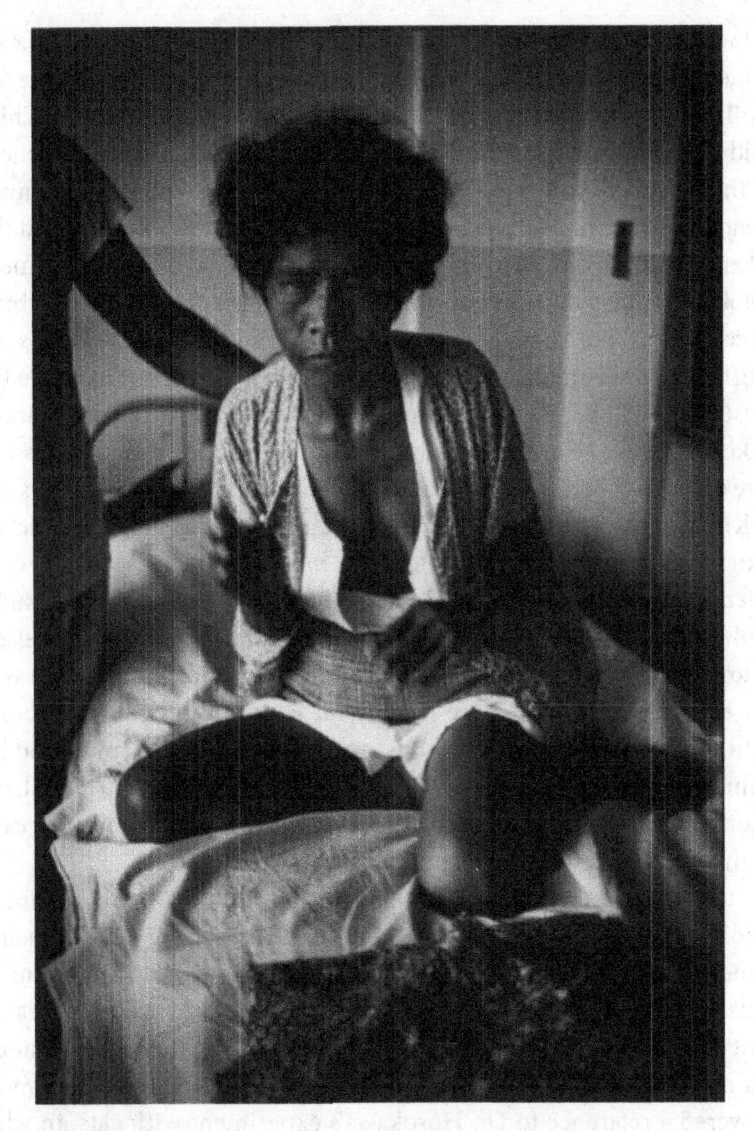

The seemingly possessed Yukijo rests between fits of uncontrolled spasms.
Credit: Shisei Kuwabara, courtesy of the artist

family realized that there was nothing they could do to pull back Ishimure from the patients. Ishimure and her husband, Hiroshi, agreed to bring Ishimure's younger sister to live with them to care for the household, and Hiroshi joined in Ishimure's efforts.

One of the families Ishimure visited was that of seventy-year-old Ezuno. Ezuno and his wife cared for their son, Kiyoto, and for Kiyoto's three sons, all of whom suffered from the disease, including Mokutaro, or Moku,[3] who had been born with it. Kiyoto's wife, Moku's mother, had deserted the family. While Kiyoto might have qualified for some compensation from Chisso, he did not apply because it would have meant a loss of government benefits, and the Chisso payment was not enough to live on. But the government stipend also was not nearly enough to live on, so Ezuno, his wife, and their weakened son, Kiyoto, were forced to fish periodically to make ends meet.

Moku, Kiyoto's middle son, was almost ten years old when Ishimure first came to visit old Ezuno and the family. Moku could not walk or stand, he suffered from convulsions, and he was not able to use his hands to hold chopsticks to eat. While handicapped in so many ways, Moku nevertheless was keenly aware of everything around him. This meant that he also was aware of his own body and his inability to control it. When Ezuno, his wife, and Kiyoto had to go fishing, and the remaining grandsons were in school, Moku was left alone at home to fend for himself. Of course, Moku could not fend at all. When Ezuno came home he knew right away if Moku had soiled himself; the shame on Moku's face broke through whatever else he was unable to communicate. He would stare at his grandfather with wide, open eyes whose sadness bore heavily on Ezuno. But Moku stubbornly taught himself how to hammer nails, and his grandfather would find him hammering nails into a wall or some loose board. Moku's blistered hands demonstrated the determination he possessed. It pained Ezuno to realize that there would be no one to care for Moku when he died.

In the struggle to gain recognition for the victims, Michiko Ishimure was joined by Jun Ui, then a young scientist and budding environmentalist, and a photographer, Shisei Kuwabara. The two were documenting the effects of the poisoning on patients throughout the Minamata area. On a visit to the Chisso Company Hospital in 1962, Ui noticed by chance a document marked "Confidential." Undeterred, Ui read the document and discovered a reference to Dr. Hosokawa's experiment with cats, in which effluent from the acetaldehyde plant, containing ten parts per million of mercury, produced symptoms in the cats that mimicked those in the patients. Kuwabara surreptitiously photographed the document.

Ui tracked down Dr. Hosokawa, who confirmed the results of his experiments but warned Ui about the risks of publishing the story. At the time, the chemical industry and the government were exerting intense pressure to suppress any publicity that would adversely affect Japan's postwar economic growth, and many of the leading faculty at Ui's university worked for the chemical industry. In 1964 Kuwabara published his photographs

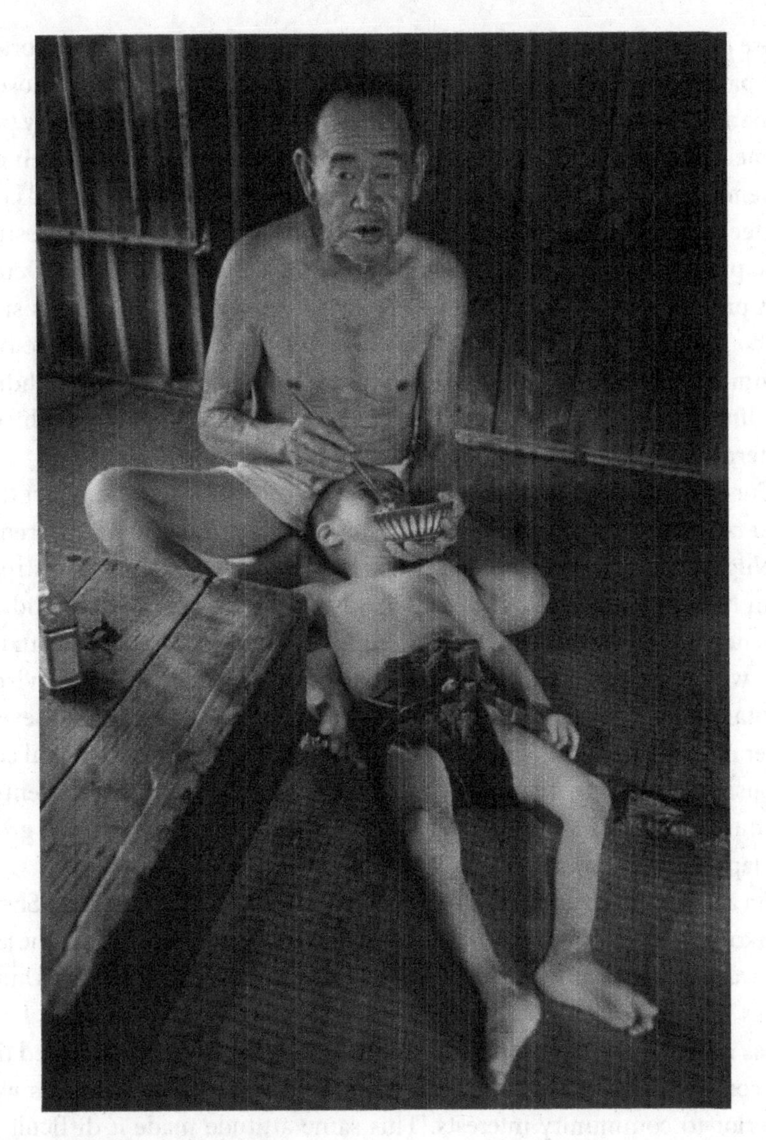

Moku is fed by his grandfather Ezuno.
Credit: Shisei Kuwabara, courtesy of the artist

of the victims, providing some of the most moving images of the impact of the disease on the victims. Ui wrote an introduction but did not reveal the secret experiments with the cats.

While Ishimure and others continued to fight on behalf of further support for the expanding numbers of victims, the company, the government, and most of the people of Minamata continued to deny the scope and cause of the disease. But another tragedy elsewhere in Japan reinvigorated the patients' efforts. In 1964 and 1965, symptoms exactly like those of Minamata Disease showed up near Niigata City. The victims, mostly poor farmers and fishermen, lived along a river near Niigata City, and their diet depended heavily on fish. As in Minamata, a government-sponsored committee was established. The committee, along with independent investigators, pinpointed the source as another chemical company, Showa Denko, that produced acetaldehyde in much the same way as Chisso, with similar consequences. Also, as in Minamata, the government suppressed the committee report and, under pressure from the trade ministers, withdrew funding for further research. The chemical company adopted Chisso's pattern of denial.

For several years, the victims in Niigata received no more support than their counterparts in Minamata. But there were several critical differences in Niigata. The chemical company was located some forty miles upstream from Niigata City, where the disease was spreading, and the city did not depend on the company as Minamata did on Chisso. Moreover, Japan had now witnessed two instances of mercury poisoning as a result of environmental pollution from the chemical industry. These instances and several other notorious acts of pollution eventually raised the environmental consciousness of the general public. The incidents also attracted the attention of young, socially committed students and lawyers, often from the growing Japanese Communist Party.

As a result of these converging forces, a lawsuit was filed against Showa Denko on behalf of the victims of the disease in Niigata. The filing of the lawsuit was an extraordinary event. While in some cultures, such as the United States, filing a lawsuit is a common step taken to resolve a dispute, in Japan it was rare and disfavored. To sue for personal compensation indicated that the community was not functioning properly, that individual rights were superior to community interests. This same attitude made it difficult for the victims in Minamata to be identified as deserving of special treatment. Thus, against all odds and tradition, the Niigata victims sued in 1968.

The Niigata lawsuit gave hope to the victims in Minamata that something might still be done to force Chisso to accept its responsibility. Ishimure published articles in the mid-1960s, and it was her writing—poetic and empathetic—that drew increasing attention to Minamata Disease and the struggles of the victims. Around this time, after the Niigata outbreak, Ui decided that he could no longer withhold information about Hosokawa's experiments. Working with Ishimure and her husband, Ui organized his research and published his findings in a limited distribution journal under an anonymous name.

One problem that remained intractable for the victims was the process of verifying whether a person was suffering from Minamata Disease. Since the initial settlement in December 1959, a procedure had been set up to certify any new individuals who claimed to be afflicted, and a Council for the Verification of Minamata Disease was formed. The process was slow and frustrating. The screening council was heavily influenced by Chisso, and its members treated the patients as greedy, unworthy supplicants. Most applicants were rejected. In fact, between 1959 and 1968 only 32 individuals were certified as having the disease, and almost half of these were congenital cases. This number was in addition to the close to 80 patients previously recognized, for a total of 111 patients, 42 of whom had already died.

Certification meant that the individual was qualified for the sum that Chisso agreed to pay in 1959. But it also meant that the government would withdraw whatever benefits it was providing to the individual, and the community at large would dispense its scorn on the individual. Members of the community saw any increase in officially recognized patients as a threat to their jobs, to the city's economic base, and to any hopes of reviving a cottage fishing industry. As one member of the Minamata community remarked, "Whose life is more important, that of one hundred eleven Minamata Disease patients or that of fifty-five thousand townspeople." Without any official recognition of Chisso's responsibility for the disease, certification remained complicated.

In 1968, the national government enacted the Pollution Victims Relief Law, which provided for the central government to take over the certification process for Minamata Disease. Contrary to expectation, the patients did not fare much better after the law was enacted. The criteria for verification remained constricted and the number of approved patients remained low.

In 1966 Chisso finally controlled the discharge of mercury from its plant through the installation of a new circulation system. In 1968 Chisso ceased production of acetaldehyde, not because of any concern for the environmental damage it was doing, nor because of the devastation that it was wreaking on the people of Minamata, but simply because new technology had made the product obsolete. Only after Chisso ceased production did the government officially conclude that Chisso's contaminated wastewater had been the cause of Minamata Disease.

However begrudging this official recognition must have seemed to the victims of the disease, it justified their long struggle to resolve the issue. Chisso, however, still refused to acknowledge its responsibility. Shortly after government recognition of the disease, patients once again held a sit-in at Chisso's plant, demanding fair compensation, not the paltry condolence that had been doled out ten years earlier. Chisso indicated that it would not engage in direct negotiations with the patients. It would instead agree only to binding mediation with the local government that would set a rate of payment for the patients, which they must accept without further complaint. The offer of mediation caused a split in the certified victims group.

Some fifty families, all of whom had been certified as having the disease, agreed to binding mediation, which was viewed as the traditional method for resolving disputes. After about a year of mediation, each living certified patient was provided a lump sum payment of ¥1.9 million ($5,515 in 1968), a maximum annuity of ¥180,000 ($155), and a medical allowance. Each family of a dead certified patient received a maximum sum of ¥3 million ($11,100). Thirty families who were certified filed a lawsuit against Chisso in June 1969, modeled on the Niigata litigation. The lawsuit brought more antagonism from the community. The victims were seen as declaring themselves worthy of special consideration and attention, and as unwilling to endure their suffering in passive silence.

The patients now at least had the support of some of the Chisso workers. During the earlier protests by the patients in front of the Chisso plant, the workers' union had been largely hostile. But in 1962 a contentious strike had split the union into two factions. One faction sided with the patients' efforts and even issued a "Shame Declaration" in 1968, apologizing for its earlier actions in defending Chisso's conduct and condemning the patients. Their support took an even more valuable turn as members of the faction secretly released confidential Chisso documents to the litigating patients.

The litigation group also received support from Ishimure. Ishimure's book, *Paradise in the Sea of Sorrow,* was published in January 1969 and it movingly depicted the effects of the mercury poisoning on the ordinary villagers of Minamata. The non-victim community in Minamata saw the book as a threat to their reputation and financial security, but throughout the rest of Japan, the reaction was positive. The book received widespread admiration, capturing a major prize for nonfiction, and fueled public interest in the plight of the victims. Ishimure refused the prize, claiming that she did not want to gain personally from the suffering of the victims. She also did not want to subject her family to further abuse on the pretext that she was writing about the victims for personal profit.

Ishimure's writing also attracted the attention of the famous American photographer W. Eugene Smith. Smith and his wife Aileen stayed in Minamata for three years, photographing victims and documenting the disease. They lived simply with a family of one of the early victims, sharing part of the modest house and eventually renting a barn for a darkroom and workspace. Aileen photographed along with Eugene, acted as interpreter, undertook the necessary research, and shared in the writing, layout, and printing of photographs for the project. Smith could speak only a few words of Japanese, but he went everywhere, observing and photographing the victims and their families. He was so ubiquitous, with cameras and lenses hanging all over him, that the locals called him "the camera store operator." As omnipresent as he was, Smith was also patient, never intruding, always waiting for that perfect moment, the perfect photograph.

Smith was particularly drawn to one young patient, Tomoko Kamimura, and her family. Tomoko was born with the poison and was limited in speech and bodily control, at best capable of uttering, "Ah, ah." Tomoko's parents refused to hospitalize her and cared for her at home. Tomoko was seen and treated as a special gift, rather than as a burden. The other children in the family did not have the disease, so it was thought that Tomoko carried the full vengeance of the disease in order to spare the others.

Tomoko became one of Smith's favorite subjects. He and Aileen lived near Tomoko's family; they walked by frequently, even babysitting for Tomoko while her parents attended rallies and protests. Smith's photo of Tomoko being bathed by her mother—a modern Pietà—remains one of the most compelling photographs of the twentieth century.[4]

Tomoko was certified, but many others were turned away. Once denied the status, there was little the victims could do. However, one victim's son, Teruo Kawamoto, challenged the government's system. Kawamoto cared for his father throughout his illness, and he was angry that the bureaucracy had denied his father the recognition that he deserved. Even after his father's death, Kawamoto pursued his father's claim for certification. When told that there was no way of proving the claim since his father had died, Kawamoto dug up his father's corpse and delivered it to the hospital, demanding that an autopsy be performed to prove that his father had died of Minamata Disease.

Kawamoto too began to feel the effects of Minamata Disease, but he resolved not to go quietly into the night. Together with Ishimure, he waged a campaign to widen the group of those eligible for certification. In August 1971, the recently formed Environment Agency overruled the decisions of the local board to deny certification and recommended that much less restrictive criteria be applied to the victims. As a result, Kawamoto and the others who had appealed were certified in October 1971, and there were an additional 538 individuals certified between 1971 and 1973.

Even after these changes, the certification process remained unsatisfying to Kawamoto and the other victims. Being certified only made Kawamoto and the others eligible for the small sum specified in Chisso's condolence. It was not the size of the condolence, however, that disturbed the victims, but the nature of the condolence. The compensation was dispensed through a third party, which allowed Chisso to distance itself from the victims. Even the litigation depended on an intercession from a third party, the court, to force Chisso to accept responsibility. During the ongoing trial, only Chisso's lawyers attended, not Chisso's key staff. The victims could tell their stories to the judge, but there was no opportunity for the victims to confront those responsible.

Kawamoto wanted direct, face-to-face negotiations with Chisso, particularly its president. He felt that only by engaging Chisso face-to-face could the company be forced to acknowledge its responsibility for the suffering it had caused the victims and their families.

In November 1971, Kawamoto organized a direct negotiations group. Their efforts were fueled by the victory in the Niigata litigation where, in September 1971, the court found in favor of the victims and found the chemical company negligent. Ishimure, as ever, worked with Kawamoto to organize another sit-in tent in front of Chisso's plant in Minamata. Ishimure and Kawamoto also coordinated their efforts with the court group. But Ishimure knew from her experience with the first sit-in, in 1959, that the victims were in danger of being marginalized by the wider Minamata community and were being pressured to submit their claims to third-party mediation. So Ishimure and Kawamoto decided to expand the protests to Chisso's headquarters in Tokyo, as well as maintaining the protests in Minamata.

On December 7, 1971, Kawamoto and Ishimure, along with several hundred supporters, appeared at Chisso headquarters in Tokyo and presented their demands for direct negotiations. The group demanded ¥30 million ($90,000) for each patient as well as an apology from Chisso. For many, an apology from Chisso and its acceptance of responsibility for the problem, not financial compensation, was the primary goal of their efforts.

The next day they returned to get Chisso's response. They met with Kenichi Shimada, Chisso's president. For thirteen hours, the victims spoke of their suffering and the suffering of their children, and pleaded with Shimada to intercede, to accept responsibility, and to make them whole again. Some were shy and reluctant; others were angry. Kawamoto was impassioned. He cried when he spoke of his father dying alone, in a mental hospital ward, without any recognition of what had been done to him. He produced a razor and pleaded with Shimada for the two of them to cut their fingers and seal in blood an agreement to settle things. The experience was too much for Shimada. He collapsed and was carried out on a stretcher. While recovering, Shimada drafted a memorandum suggesting that the company turn the plant over to the victims and their supporters as compensation for their suffering. Given the culture of Chisso, nothing could come of such an idea.

With the talks suspended, the protesters camped out in Chisso's offices to wait until Shimada agreed to meet with them again. After several weeks, Chisso workers forcefully threw Kawamoto, Ishimure, and the others onto the street. Undeterred, the victims set up a tent outside, as they had previously done at the Minamata plant, to continue the protest. The forcible ejection received widespread publicity, and the ranks of the victims' supporters grew.

Kawamoto felt that the Chisso union should not be acting as Chisso's enforcers. Most of the workers involved came from Chisso's plant in Goi, so Kawamoto decided to visit the leader of the union in Goi to encourage them to desist. On January 7, 1972, Kawamoto, some supporters, and members of the press, including Eugene and Aileen Smith, traveled to Goi for a meeting between Kawamoto and the head of the union. When they arrived, Kawamoto was informed that the union leader was in Tokyo and not available for any meeting. Kawamoto tried to deliver a set of demands but a group of Chisso workers charged and beat them, particularly singling out Smith. Smith was kicked in the groin, and then several of the Chisso workers picked him up and slammed his head against concrete, knocking him unconscious.

Smith developed recurrent dizzy spells, a constant pain in his left eye, and blurred vision in his right eye. Sometimes he fainted if he tried to lift his hands to use the camera. When whiskey was not enough, Smith got painkillers at the local hospital, and occasionally traveled to Tokyo for treatment by a chiropractor. Eventually, he went to New York on several occasions to get treatment, always returning to Minamata where Aileen had remained to work on their book of Minamata photographs.

The attacks by the Chisso workers aroused a public outcry. Yet, despite the critical climate and pressure on Chisso from the Environment Agency, further talks were fruitless. Chisso even had Kawamoto criminally charged for attacking Chisso workers at one of the confrontations at the Tokyo headquarters.

While the trial of the victims' claims against Chisso was proceeding, Dr. Hosokawa was dying of lung cancer in a Tokyo hospital. Dr. Hosokawa had been loyal to Chisso even after retirement, but he was deeply troubled by what he knew about the discovery of the cause of the disease. When the disease broke out in Niigata, Jun Ui asked Hosokawa to visit Niigata and confirm the disease. Hosokawa did so, and when he returned he asked Chisso to release him from his obligation to remain silent about the cat experiments. Chisso refused. Ishimure visited Dr. Hosokawa in the hospital and he asked about the children born with the disease—were they growing, were they feeling better? Then he took Ishimure's hand and held it to his chest and asked her if she could feel a lump. That was the cancer, he told her. There was more on Dr. Hosokawa's chest. He spoke about the need for repentance, not only for himself but also for the Chisso managers. He recognized that the clock could not be turned back nor the damage undone. But he felt that if the company officials did not move quickly to repent, they could not be redeemed and the evil would grow and worsen.

At the trial Hosokawa provided a deathbed testimony that shocked the wider public. He testified about his discovery of the link between the acetaldehyde production wastewater and the disease, back in 1959 when Chisso was settling cheaply with the patients; he also testified that Chisso suppressed the experiments and destruction of the cats. His revelations exposed Chisso's conduct of continuing to discharge mercury from the acetaldehyde process knowing that its wastewater was likely the cause of the devastating disease. Hosokawa died several months after his testimony, and the victims erected a small shrine to him outside their protest tent.

The trial verdict was delivered on March 20, 1973. Chisso was excoriated. The Kumamoto District Court found Chisso grossly negligent, stating: "no plant can be permitted to infringe on and run at the sacrifice of the lives and health of the regional residents." The court also nullified the solatium/condolence agreement of 1959, holding that Chisso took advantage of the victims, and that it continued to discharge wastes contaminated with mercury even after it knew of the link between its wastes and the disease. In response to Chisso's defense that it could not foresee the harm it caused, the court replied that the victims were not guinea pigs. The court awarded ¥937 million ($3.6 million) to the thirty families. These awards were substantially higher than the awards granted through mediation.

Following the reading of the verdict in court, the victims gathered outside the courthouse, carrying photographs of their loved ones who were afflicted. The condemnation of Chisso was an important victory, but the suffering of those victims remained. Tomoko Kamimura's mother held her outside the courthouse and cried out that her child had been priced at 18 million yen but that she would never be normal. To her mother's remark, Tomoko added, "Ah, ah."

The judgment of the Kumamoto District Court was an end but not the conclusion. The litigation victims joined forces with Kawamoto's group and demanded face-to-face negotiations with President Shimada for medical expenses and annuities, in addition to the lump-sum death benefit payments awarded by the court. The groups insisted that all victims receive the same compensation. Most of all, they wanted an apology from Chisso.

Chisso agreed to face-to-face negotiations. Once again, Kawamoto, who had organized the earlier negotiations and the confrontation with the union in Goi, sat like a specter on top of the table, cross-legged, staring directly into Shimada's face as they spoke. At one point, Kawamoto asked Shimada if he had any religion. Shimada said he was a Zen Buddhist and that he kept a small shrine with the names of all the victims and prayed there.

Yet the talks dragged on. Frustrated with the delays, one of the victims got up, shaking, and smashed an ashtray on the negotiating table, cutting his wrists with the jagged edge. Shimada could remain silent no longer, and shouted, "We'll pay, we'll pay." Chisso finally agreed to a global negotiation with all the groups and a settlement was reached in July 1973. Under the settlement, each of the certified patients received a lump sum of ¥16–18 million ($51,000–$59,000), plus a lifetime monthly pension. In addition, Chisso agreed to pay for medical and economic aid, and 65 percent of the cost of cleaning up the bay. Perhaps most critical for the victims, Chisso apologized to them and to society. In addition, criminal charges were brought against a former Chisso president and plant manager for manslaughter. The charges were reduced to professional negligence, the two men were found guilty, and the Japanese Supreme Court upheld the convictions.

With the determination of Chisso's responsibility settled, cleanup of the contaminated bay proceeded. A three-mile-long net was stretched across the bay to prevent fish from leaving the contaminated area. Then the government excavated approximately 2 million cubic yards of mercury-contaminated sediment from the bottom of the bay and created a landfill parkland of about 143 acres. By 1997, the remediation was complete and the net was removed. The government declared that fishing in Minamata was once again safe.

Those who were exposed to the mercury poisoning did not feel so protected by their governments. Many felt that the regional and national governments were as culpable as Chisso. Several lawsuits were brought against the governments, charging them with failing to investigate the poisoning and failing to determine early on who had eaten the mercury-contaminated fish and the extent of their injuries and symptoms. In the period after World War II, when economic expansion depended on the chemical industry, governments supported Chisso in its efforts to avoid responsibility and held environmental and health concerns secondary to their economic goals. Victims pressed one case all the way to the Japanese Supreme Court. In October 2004, the court ruled in favor of the victims. It found that the national and prefecture governments had failed in their responsibility to identify the source of the mercury poisoning and to stop it, and that the governments had exacerbated the suffering of the people.

Yukijo remained hospitalized on an island off the coast, where she had moments of lucidity and often played music. She died in the early 1970s in her late fifties. Kawamoto remained actively involved throughout his life, helping uncertified patients, working to establish the Minamata Disease Center Soshisha, an important resource and support center, and becoming a city councilman. He died in 1999 at the age of 67.

Following the verdict, Eugene and Aileen Smith remained in Minamata to finish their book. In early 1974 they returned to the United States and, in 1975, published *Minamata* to wide acclaim. The photo of Tomoko in her bath was among Smith's last great photographs. Tomoko died in 1977 at the age of 21; Eugene Smith died in 1978, following a stroke. At a private graveside service, a telegram arrived from Minamata that read, "We come upon the unexpected news of your death and profoundly cannot endure our grief. Your history is our courage itself. We pledge our inheritance of the mighty footsteps you left behind at Minamata." Aileen Smith returned to Japan and now directs Green Action, an important antinuclear organization in Kyoto.

Michiko Ishimure continues to provide comfort and courage for the patients, and to write about their suffering. She has written a Noh play, *Shiranui* (Sea Fire), about the sea around Minamata and the tragedy it released. The play was performed in 2004 on the area of the sea that was converted to land to protect people from further exposure to the poisonous mercury.

Moku persists. His grandfather, Ezuno, died, as did his grandmother, but he is cared for at Meisui-en, a facility established for Minamata Disease patients. While he remains in a wheelchair, and his physical movements and speech are severely limited, Moku has become an accomplished photographer. He was trained by Kuwabara, the photographer who, with Jun Ui, discovered the secret document of Dr. Hosokawa's cat experiments and who produced some of the most moving photographs of Minamata patients. Moku's photographs have been published and exhibited. His pain and persistence embody the courage and struggle of those deeply affected by Minamata Disease.

Chisso's economic dominance allowed it to get away with denying responsibility for those suffering from the poisoning, to pay a pittance to the early victims, and to continue to discharge the toxic wastes even after it knew that its wastes were causing that suffering. Externally, the company aligned itself with commercial interests within the government that turned a blind eye to the company's culpability and to the victims' suffering.

Under the circumstances, and given the importance within Japanese culture of preserving the cohesion of the community, it is remarkable that individual victims had the courage to fight the company, the government agencies, and their neighbors and community. In the long run, the victims' tenacity served them well as they formed various victim and support groups, organized sit-ins at Chisso facilities and demonstrations in Minamata and Tokyo, and initiated lawsuits against Chisso and the governments. It is likely that the tough independence of the fishing people and comfort from family members allowed them to endure. The struggle was certainly blessed by the moving writing of Michiko Ishimure and the compelling photographs of Shisei Kuwabara and W. Eugene and Aileen Smith.

LONDON, ENGLAND
1952

The most unusual fact about the London Fog of 1952 was not that some four thousand people died from it—one of the largest numbers of people killed by any environmental disaster—but that no one seemed to recognize that it was happening. For four days the fog was so thick that traveling throughout the city was almost impossible, but few realized just how deadly it was. After all, London had been notorious for its fog for a long time—romantic notions were even attached to it. For the residents of London the fog was a frequent, if unwelcome, guest.

In 1952 Londoners were relying heavily on soft, bituminous coal for fuel. The soft coal was cheap, in part because of the low cost of shipping it by sea from Newcastle, but it had a higher sulfur and nitrogen oxide content than the harder anthracite coal used in Wales and Scotland. The smoke it emitted was tarry and full of hydrocarbons.

When carbon particles of soot from coal-fire emissions combine with particles of water, fog becomes smog. The soot and water combination is not transparent to light, and as the fog thickens, light is prevented from penetrating the foggy air. No only does this cause limited visibility, but a breath of this air carries with it carbon particles and other dangerous substances.

Certain weather conditions, particularly temperature inversions, aggravate fog. Usually the air near the ground is warmer than the air higher up, and the warm air rises and mixes with the cooler air. Occasionally this relationship is inverted with the colder air remaining close to the ground and the warmer air above it, trapping the colder air on the ground. If there is little or no wind, the air becomes stagnant and anything in that air, such as soot, remains suspended.

During the nineteenth century, clean-air advocates attempted to address the emissions from factories and other businesses that contributed much of the soot. Eventually, they met with some success as legislation was passed making it a nuisance for a chimney to emit black smoke from a commercial establishment. Yet enforcement was difficult and sporadic, especially with regard to proving what constituted black smoke.

The smoke from domestic hearths remained uncontrolled. One problem in regulating domestic sources was the lack of alternative smokeless fuel supplies. Just as difficult an obstacle was the English fascination with a "pokeable" open fire. It was considered a national entitlement to make an open-hearth fire, and it was a sign of affluence, as well as of hospitality, to have a blazing hearth. By the first few decades of the twentieth century, those pokeable domestic fires, along with industrial emissions, dumped some seventy-six thousand tons of soot on London each year, the equivalent of about 650 tons for every square mile. About two-thirds of the smoke in London came from domestic fires. During World War II, the government even actively encouraged businesses to pollute as military authorities thought the smoke would serve as camouflage and make it more difficult for the German bombers to see their targets. Even after the war, the fog remained an accepted aspect of living in London.

Though the typical winter climate was cold, damp air with some clearing spells, followed by fog or rain or snow, the fog dominated London during the first week of December 1952. On Thursday evening, December 4, a high-pressure system settled over London, and a temperature inversion trapped in the fog throughout the area. By Friday morning, tons of carbon particulate and sulfur dioxide poured out of millions of domestic coal fires and industrial plants into the still, foggy air over London. The temperature inversion prevented the dispersal of the fog into the upper air and trapped the smoke and other pollutants at ground level. Smoke that escaped from the tall stacks of the manufacturing plants fell to the ground rather than rising into the air.

On Friday, the fog and smoke covered much of London. A visitor staying in a warm, dry hotel with nothing to do might have found the fog on that first full day to be charming. Those who had to go to work did not. In the morning, people could see the outlines of buildings from a distance of only seventy to eighty yards; by noon, the large sculptural figure

Londoners carry on as the fog descends.
Credit: ©Henri Cartier-Bresson/Magnum Photos

atop Nelson's Column on Trafalgar Square was barely visible. Around the Houses of Parliament, visibility was limited to a dozen yards. By that time, streetlamps had to be lit. With visibility along the Thames at zero, the Port of London was forced to close. Airports also closed. As the day wore on, travel became increasingly difficult. Buses everywhere in London experienced serious delays.

The color of the fog was not the usual gray, but black, or at times yellow. As evening fell, the Christmas lights in store windows looked eerily suspended in open air since the stores themselves could not be seen from a short distance. Flares were placed at intersections for the vehicles still on the streets. People groped along buildings, stumbled over curbs and each other, and when they arrived home found they were covered with soot.

More disturbing than the impaired visibility was the difficulty in breathing, especially for older people and those with bronchitis. The smell of sulfur permeated the air. Noses stung, throats felt tight, and people coughed up blackness.

When Londoners awoke on Saturday morning, the sixth, the fog was yellow and thick. It extended over an area of one thousand square miles. Very few buses operated. At one point, seventeen buses formed a caravan to try to find their way back to the garage. The famous red double-decker buses inched along, bumper-to-bumper, with conductors leading the way by walking in front with flares, shouting directions. Ambulances traveled the same way. The fog infiltrated the tube stations. At one station, a bride and groom were waiting for a train to take them to their reception, since they had to abandon street-level transport. The bride's wedding gown was black from the soot in the air. By Saturday evening, the fog followed people inside, through open doors, down chimneys, even through cracks in walls, floors, and windows. Hospitals began to fill up. Yet by late Saturday, the BBC was reporting only that the fog might persist. No emergency had been declared.

By Sunday, everything was blackened, inside as well as outside. Visibility remained at a few yards. Ambulances ran out of flares. With so many patients needing assistance, the ambulances began to carry several on each trip to the hospital. On one trip, an ambulance that had been dispatched to carry four patients to a hospital ended up taking them all to the mortuary instead.

The elderly and the sick, especially those living alone, were increasingly isolated during the fog. They could not get out, and if they did, they could hardly breathe. As one elderly patient described it:

> It makes you feel certain that you're going to die, that death is surely coming for you, partly because of your difficulty in breathing and partly because of the fierce pain in your throat and lungs...and adding to your terror is the sight of the fog, when you see it there all around you, like some kind of gray, obscene animal, outside your window, drifting, floating, almost looking in at you, as though it were waiting there to claim you, to seize you, to choke you...to squeeze the breath, the very life out of your body.[1]

On Monday, the fourth full day of the fog, forecasts suggested that the fog might be lifting, but they were wrong. While the air west of London cleared somewhat, conditions over the city remained stagnant. Vehicles were abandoned all over the city. In the Underground, the only viable means of transportation, long lines formed at the ticket booths. A performance of Verdi's *La Traviata* was canceled after the first act because fog inside the theater made the stage invisible. Early in the evening, the BBC broadcast that the fog was dirtier than usual and that coal-burning domestic fires were partly to blame. The item was deleted, however, from later broadcasts.

Finally, early on Tuesday morning, December 9, a slight breeze blew across London and the fog began to lift. By 9:00 AM, the Thames cleared of fog, and the port reopened. More than one hundred ships waited to leave the port; over two hundred ships waited to get in. The city began to breathe more easily.

The disruption of travel and sporting events dominated coverage in the papers. In the days following the lifting of the fog, letters to the *Times* debated only the economic benefits of electric versus coal heat. Few recognized the environmental or health hazards of the fog.

Soon, however, its human costs became visible. Doctors reported significant increases in respiratory disorders over previous winters. During the fog, hospitals around the city experienced a rise in emergency admissions, especially for respiratory ailments. The hospitals remained filled for days even after the clearing.

By mid-December the papers reported that as many as one thousand Londoners had died as a result of the fog. Questions were raised in Parliament, and the health minister responded that the deaths attributable to the fog may have been as many as three thousand. Smoke abatement advocates demanded an investigation. The government resisted. Harold Macmillan, then a cabinet minister, remarked in private that they should form a committee that would do little but would appear busy, in an effort to calm the public. It was not enough and the air pollution committee in Parliament, named after its chairman, Sir Hugh Beaver, addressed the matter with all due seriousness. The Beaver Committee castigated both the local and national governments for failing to take preventive measures to protect the public. They also laid blame on domestic consumers as the largest producers of smoke and recommended the limit of smoke from all chimneys—both industrial and domestic—the production of greater supplies of smokeless fuel, and the establishment of smokeless zones in urban areas.

In January 1954, an article in the respected *British Medical Journal* estimated that the fog had caused over 4,500 deaths. That same year, the Ministry of Health produced a report that analyzed the effects of the fog. The government recognized that throughout those early days of December the metropolis of 8.5 million people was hardly aware that a disaster was occurring. The residents were also unaware that the aftereffects had continued to affect the city for several weeks. The concentration of the dark smoke was detected at 4,500 micrograms per cubic meter and sulfur dioxide at 3,700—five to ten times that of normal levels.

The Ministry of Health concluded that there were as many as four thousand more deaths than would normally have occurred in the first three weeks of December, and that these deaths were caused by the fog, and in particular its tarry particles and sulfur oxides. The deaths were concentrated among people with preexisting respiratory or cardiac disorders and among the vulnerable, those over sixty-five years and those under one year old. The source of the contaminants was identified as irritants derived from the combustion of coal.

The report further suggested that many of those who died from the fog had already been suffering and were expected to die within a short time anyway. This concept was referred to as short-term mortality replacement or, more graphically, "harvesting." But when the number of deaths over the following weeks was analyzed, it was determined that there was no drop in the number of deaths. This led many to believe that those who died during and immediately after the fog were not "harvested" but killed.

Only after further agitation by antismoke factions and other civic groups did the government address the issue through the Clean Air Act of 1956. For the first time, regulations subjected domestic coal fires to controls, established an objective measurement for what constituted dark smoke, and empowered local governments to establish smokeless zones in their areas.

The British Clean Air Act of 1956, implemented slowly over a decade, significantly reduced smoke caused by domestic fires. For example, when smog covered London in December 1975, the peak concentration of smoke and sulfur dioxide did not exceed 800 micrograms per cubic meter and 1,200 micrograms per cubic meter, respectively, or less than 20–30 percent of peak levels during the 1952 fog. Besides prompting the Clean Air Act, the 1952 fog served as a catalyst for the study of diseases and deaths attributed to air pollution, leading to regulation of ambient air quality in many other countries, including the United States. Studies over the past fifty years have led to an increased understanding of how soot, fog, and particulate matter affect populations, especially in demonstrating the correlation between high concentrations of particulate matter and respiratory diseases and deaths. Based on more advanced research techniques, a recent reassessment of the effects of the 1952 fog estimates that as many as seven thousand to twelve thousand deaths, not four thousand, resulted from the fog.

Though Londoners moved away from soft, high-carbon coal to smokeless fuels, they also grew more reliant on cars for transportation. While catalytic converters have reduced emissions per vehicle, the number of vehicles in London has grown so significantly that vehicular emissions are now the primary threat to the health and environment. Londoners may rely less on dirty coal fires, but their dirty and dangerous oil-fueled cars are quickly becoming as great a problem.

WINDSCALE, ENGLAND 1957

Britain aspired to become a nuclear power after the Second World War. At Hiroshima and Nagasaki the Americans had demonstrated their capabilities for nuclear power. While the Americans and the British were close allies, in the Atomic Energy Act of 1946 the Americans prohibited the sharing of any nuclear information with anyone outside the United States, including Britain. The Soviet blockade of Berlin during 1948–1949 and the detection of the first Soviet atomic bomb test in 1949 exacerbated the military insecurities felt by the British. Undeterred by the cold shoulder given to them by the Americans, the British pushed ahead, amid great secrecy, with a facility to produce plutonium, a necessary ingredient in the atomic bomb. The British nuclear facility, originally known as Windscale and now known as Sellafield, was located in Cumberland on the Irish Sea. Construction began in 1947, and operations at the plant began in 1950. Using plutonium from Windscale, the first British atomic device was detonated in 1952 off the coast of Australia. Windscale enjoyed an auspicious start.

The equipment at Windscale included two air-cooled plutonium production reactors that were contained within large graphite blocks. Fuel cartridges with uranium filled the thousands of channels that ran through the blocks. Neutrons bombarded the uranium to create heat and plutonium through fission. The plutonium was extracted from the cartridges at another plant on the Windscale complex.

An unwanted byproduct of the process includes something known as "Wigner" energy; neutron radiation causes material like graphite to undergo changes in its physical properties. During this process graphite actually swells, its thermal and electrical conductivity decreases, and it retains thermal—Wigner—energy. Unless this Wigner energy is released under controlled conditions, heat builds up and can lead to an explosion. If the reactor is heated up slowly to about 392°F, the graphite returns to its normal condition and the Wigner energy is safely released. The process is slow, time consuming, and disruptive of plutonium production schedules. At Windscale, the procedure to release the Wigner energy kept getting pushed back with longer intervals. The longer the delay, the more Wigner energy was stored in the reactor, and the greater the potential for danger.

Reactor No. 1 was due to have its Wigner energy released in the fall of 1957, a time when plutonium production schedules were particularly tight because of another bomb test scheduled for June 1958. Since it had been done on eight previous occasions, no particular problems were anticipated. The procedure was initiated on Monday, October 7, 1957.

The heating of the reactor began Monday evening and stopped the following morning. Normally the Wigner release would have continued on its own momentum, but on this occasion falling temperatures indicated that the release was slowing. The operators were unaware at the time that the thermocouples, the instruments that measured the temperature, had been placed in such a way as to accurately measure the reactor's temperature only when it was operating normally. Some of the thermocouples indicated one temperature where they were located, while the instruments in an adjacent area of the reactor indicated that the temperature was one hundred degrees hotter. Believing that the temperature was falling and that the Wigner release might not complete itself, thereby creating a danger of spontaneous combustion at a later date, the operators increased the heating of the reactor again. The temperature rose unexpectedly fast, so the power was reduced. This raising and lowering of the temperature, through the use of dampers, continued throughout Tuesday and Wednesday.

Unknown to the operators, inside the reactor the aluminum that encased the uranium had cracked, exposing the uranium to the outside air and causing it to burn. By midday on Thursday, October 10, the dampers were opened once again. At this point a noticeable increase in radioactivity was detected in the stack, and it was clear that the temperature of the core was rising quickly. The operators were unsure as to what was going on, so they reported their concerns to managers.

Around this time, the health and safety manager received reports of increased radioactivity both in the stacks and in air measurements on the ground half a mile from the reactor building. Theorizing that perhaps a fuel cartridge had burst, monitoring equipment was dispatched south along the coast to Seascale, a town that was believed to be downwind from the stacks.

The health and safety manager conferred with the assistant works manager. They were beginning to distrust the reactor's instrumentation, and it soon became clear that no one knew what was going on inside the reactor. The only thing to do was to observe the reactor directly. Several managers opened plugs in the reactor and peered inside. Expecting to see black fuel elements, the managers instead saw that the fuel elements were glowing and white hot, which meant that something—the uranium or the graphite, or perhaps both—was on fire.

Workers were ordered to don full protective gear and push the fuel cartridges out from the core to try to stem the fire. At first, they tried to remove the burning fuel rods with sledgehammers and scaffolding poles to slow down the "thermal runaway." But the cartridges were already so damaged by fire that they would not budge. Managers gathered to devise a plan to deal with the crisis. Some speculated that if the temperature kept rising, the entire nuclear reactor would catch fire and the radioactive components of the core would discharge into the atmosphere and over the countryside. It was a terrifying prospect that was becoming all too real.

By early Friday morning, managers notified local authorities of a fire at the plant and set in motion emergency procedures, including the preparation of buses to evacuate the neighborhood if necessary. Workers at the Windscale site were told to stay indoors and to wear respirators if they went outside, but no one warned anyone in the surrounding area. Farmers on Saturday observed dark orange smoke coming from the stacks, but they went about their usual chores. Students from the Seascale prep school played on the banks of a river several hundred yards away from the stacks. The earliest BBC newsreels reported that "The Atomic Energy Authority have announced that some uranium cartridges in the center of the atomic pile at Windscale became overheated yesterday."

Carbon dioxide was piped into the core to smother the fire, but to no avail. There was no other choice but to try flooding the core with water, destroying it beyond repair. No one knew if flooding would work, or if it

Several days after the accident, all milk within eighty square miles of the facility was confiscated. Eventually, residents within two hundred square miles of Windscale were warned not to drink milk.

Credit: Photography by Robert Del Tredici, courtesy of the artist

would cause a massive nuclear explosion. At 9:00 AM, Friday, October 11, operators turned on the water, a large hiss was heard, and several minutes passed. Anxious silence among the operators followed, but there was no explosion. Within an hour the fire was quenched. As a further precaution, water was poured on the core for another day. Only then did the operators notify local authorities that the fire was under control.

Both the heat from the burning uranium and the steam from pouring water on the core released radioactivity into the air. After the fire was controlled, officials were able to assess who was at risk and to determine what environmental damage had been done. They sampled vegetation, soil, grass, and foodstuffs, including milk, which presented special concerns. The radioactivity contained iodine-131, which is absorbed by grass, eaten by dairy cows grazing in the area, and then transferred to humans, including infants, through the milk. Officials decided that milk with iodine-131 in excess of 0.1 microcurie per liter was unsafe, and by Saturday, October 13, milk samples contained levels of iodine-131 between 0.4 and 0.8 microcurie per liter. All milk within eighty square miles was confiscated.

The initial assumption was that the ground-level weather vane indicated the direction of any radioactive contaminants escaping from the reactor. On Thursday, the wind at ground level came from the northeast, spreading radioactivity southwest toward Seascale and out over the Irish Sea. What wasn't taken into consideration at first, however, was that southwest winds higher in the atmosphere spread contaminants north and east of the site. Further complicating matters, a cold front blew in from the Irish Sea on Friday, increasing the wind from the northwest and spreading radioactive contaminants southeast over England, and then over Belgium, Germany, and Norway. After a fuller analysis of these wind patterns, authorities extended the milk ban on October 15 to an area of 200 square miles surrounding Windscale.

Despite the delays in notifying the public, there were no protests or expressions of anger toward the government. The Cumberland area was economically depressed in the postwar period, and Windscale represented jobs and economic security. Besides, in 1957, organized opposition to nuclear facilities was unknown.

Opposition would no doubt have surfaced sooner if the British government had not suppressed the critical components of a report on what actually happened at Windscale. A government-sponsored investigation, the Penney Report, identified the cause of the nuclear fire as the second nuclear heating on Tuesday, October 8, when the fuel cartridges failed from the rapid rise in temperature. The report also made numerous recommendations for improving safety at nuclear facilities in Britain. While the British Atomic Energy Authority and even the Ministry of Defense approved the report for publication, Prime Minister Harold Macmillan suppressed it. For years the British had tried to convince the Americans to lift the ban on the exchange of nuclear technology, and in the fall of 1957 the issue was before the American Congress. Macmillan visited President Dwight Eisenhower on October 23, and they issued a declaration pledging their commitment to sharing nuclear information and technology. Eisenhower promised to seek an amendment to the Atomic Energy Act to open American nuclear expertise to the British.

When Macmillan returned from Washington, the Penney Report was waiting for him. Macmillan believed that a frank discussion of what went wrong at Windscale would undermine, even destroy, his efforts to win congressional support for the amendment. Macmillan released only parts of the Penney Report, and substituted a white paper that softened the critical conclusions and recommendation sections. The full report was suppressed under Britain's secrecy law for thirty years. Only in 1988 did the British public receive a fuller account of the events at Windscale.

The story did not stop there. The government's white paper concluded that there was no immediate danger and no likely ill effects on health from the Windscale fire, except for the contaminated milk. Over the years, many have challenged this conclusion.

Various scientific and governmental bodies, between 1960 and 1990, estimated that some 20,000 curies of iodine-131 were released, along with various other radioactive contaminants, including strontium-89, strontium-90, cesium-137, and polonium. Analysis of these contaminants, a growing understanding of the pathways of exposure, and changing assumptions about the relationship between dose and effect from low-level radiation led to new assessments of the health risks resulting from Windscale. It has been concluded that some 250 cases of thyroid cancer and between 120 and 300 deaths from other cancers occurred in the United Kingdom over a forty-to-fifty-year period as a result of Windscale.

Contaminated water and contaminated milk were dumped down drains and into the sea, and the buildings at Windscale were decontaminated. Nevertheless, the ending for the Windscale story remains open. Only now are the British able to plan a cleanup of the nuclear pile, fifty years after the crisis, and the estimates are that it will take until 2025 and cost $1 billion.

Twenty years later, Americans were to get a taste of just how terrifying the prospect of nuclear meltdown could be when a place called Three Mile Island became known to the world.

SEVESO, ITALY
1976

The July weather was, as usual, hot and sticky. Nothing else was usual for the refugees that summer. With little warning the authorities had ordered them to evacuate with only the clothes on their backs and a suitcase. Most of those who were forced to leave had built their homes in their spare time, either by themselves or with the help of relatives and neighbors. Most had gardens, even small farms, attached to their houses. Now all of this was taken from them. Barbed wire fencing—nine feet high and some six miles long—was constructed around the area. Armed soldiers guarded the area—the Zone.

All of this paled in comparison to the sight of the faces of their children, swollen with pustules and running sores, and covered with black and scarlet pockmarks. The children were taken away and put in camps during the day, to protect them from the dangers in the Zone. Despite protests from the authorities, a number of pregnant women had abortions rather than risk giving birth to newborns with serious deformities because of exposure in the Zone. Their lives had become, as they said, *bruttissima*, the ugliest kind of life.

The affected area was Seveso, Italy, a small town of working-class families north of Milan. Its location was close to Italy's industrial heart yet far enough out in the Lombard countryside to attract multinational corporations.

One such company was the Industrie Chimiche Meda Societa Anonima (ICMESA), which built a plant in Meda, a town adjacent to Seveso.

Givaudan, a Swiss company established in 1898, owned ICMESA, and the large international drug manufacturer Hoffmann-La Roche—also a Swiss company—owned Givaudan.

Givaudan produced a variety of aromatic essences for the cosmetic industry; it also produced a bacteriostatic agent called hexachlorophene for surgical soaps and other toiletries. The ICMESA plant near Seveso made intermediates for further processing by Givaudan, including aromatic compounds for perfume essences and something called trichlorophenol (TCP), a component of hexachlorophene.

Hexachlorophene was originally developed as a bacteriostatic agent to stop the spread of infections and was used by physicians in scrubbing up before operations. By the late 1950s, it was being hailed as another miracle chemical ingredient and was widely used in consumer products, including soaps, deodorants, talcum powder, acne treatment, and vaginal sprays. Hospitals used it for washing newborn infants. Beginning in the late 1960s and early 1970s, however, animal studies indicated that large doses of hexachlorophene could result in paralysis and brain tissue damage. Some studies suggested that it could cause convulsions in babies. The dangers of hexachlorophene became evident in 1972 with the deaths of more than twenty French infants who had been treated in their cribs with talc that contained excessively large amounts of hexachlorophene. Subsequently, hexachlorophene was banned in the United States for all uses except surgical soaps. However, it was still widely available in other countries, including Italy.

Another danger of hexachlorophene was a deadly unwanted byproduct of trichlorophenol (TCP). When the temperature of TCP went over 200°C (392°F), tetrachlorodibenzeno-p-dioxin (TCDD), or dioxin, was accidentally formed. Dioxin had a menacing past. Givaudan knew that dioxin had been released in accidents at TCP plants in England, Germany, the Netherlands, and the United States from the late 1940s through the mid-1970s. The results were disturbing. Besides a horrid skin condition called chloracne, there were reports of deaths, cancers, and systemic poisoning among those exposed to the dioxin. An herbicide derivative of TCP was first used for military purposes by the British in Malaya in the early 1950s. It was made infamous by the Americans in Vietnam where, formulated with another herbicide, it was known as Agent Orange.

TCP was made by ICMESA by mixing tetrachlorobenzene with sodium hydroxide in an ethylene-glycol solvent. The materials were mixed in reactors, and then subjected to a distillation process to remove the ethylene-glycol solvent. On Saturday, July 10, 1976, something went terribly wrong with the process. The reactor had shut down before the distillation process was completed. Workers switched off the steam that heated the reactor and stopped mixing the chemicals. The liquid mass remained in the reactor and, even though the controls were set for cooling, the chemicals continued to heat up. The temperature likely rose to 300°C and a safety disk at the top of the reactor blew out. The boiling mixture of chemicals exploded into the open air, forming a white cloud. A foreman grabbed a gas mask and ran into the reactor room and turned on a valve to let water into the reactor's cooling system. The water reduced the heat, and the release of gaseous materials stopped within half an hour.

The explosion—a loud, screeching, hissing sound—caught the attention of the residents of Seveso and Meda. As they turned toward the sound, they saw the cloud, varyingly white and gray, heading toward and then over them. The cloud descended on them, like a fog filled with damp crystals. The fallout caused coughing and burning eyes, quickly followed by headaches, dizziness, and diarrhea. The people were accustomed to smoke and pollution from the surrounding industries, and therefore were not alarmed initially. As one resident stated, "At that time, we didn't pay much attention to it, because there was always some cloud, some leak. And although there was a terrible smell, we didn't worry about it; we ate, we collected vegetables and flowers from the garden, as if nothing had happened."[1]

The northerly wind pushed the cloud south, covering an area four miles long and a third of a mile wide within half an hour. The chemicals in the cloud dispersed over the area, hitting Seveso the hardest, but also spreading over parts of seven nearby towns, including Meda, Desio, and Cesano Maderno.

When people began to call the local police for information, no one could tell them anything. Some residents went to the ICMESA factory to find out what had happened, and were told that there was nothing to worry about. ICMESA's director of production, Paolo Paoletti, informed the mayors of Seveso and Meda that the cloud was an "aerosol mixture" with some possible toxic substances. The company took samples of the material and sent them to Switzerland for analysis. The company also requested that the local residents be advised not to eat vegetables from their gardens or fruit from their trees. The local officials, however, did not know how far or how wide the cloud had dispersed, so they warned no one.

Residents awoke on Sunday to the lingering acrid smell. Crystals, now shiny and oily, covered everything, and the cloud still hung over them.

Since the weather remained warm, the children played outside, trying to ignore the smell. Fruit and chickens and rabbits were selected from the family gardens for the Sunday evening meal. Some people suffered from headaches and swollen eyes after eating their meals. Fortunately for the residents of Seveso and Meda, on Sunday evening a strong wind from the Alps blew away the cloud that remained. But the wind only spread the cloud and the fallout farther out.

On Monday, the local health officer inspected the plant and met with the ICMESA manager. Workers complained of minor headaches, nausea, and burning sensations, but not of any serious injuries. Following the meeting, the ICMESA manager confirmed in a letter to the health officer that chlorinated phenol was the major component of the cloud but that ICMESA could not otherwise identify the substances in the vapor. The company did not provide the Italian officials with any results of the sample analyses, and there was no mention of dioxin. Comforted by the letter, the local health officer concluded that the cloud posed no risk for the area and so informed the provincial health officer in Milan.

At the plant, workers threatened to strike because they believed that the company had withheld information about the risks. They had already gone on strike several months earlier to force the company to allow a regional health inspector to enter the plant and evaluate health hazards. On Friday, they posted signs in the immediate vicinity of the factory with the following warning: "DANGER AREA. CONTAMINATED. DO NOT EAT VEGETABLES, FRUIT, OR ANIMALS THAT EAT GRASS FROM THE GROUND." The signs were alarming for residents because they learned for the first time that the fruit and animals they had been eating for a week might be contaminated.

After the explosion, Givaudan and Hoffmann-La Roche analyzed the soil and dust samples that had been collected at the factory and in the surrounding area. Dioxin is difficult to detect, and the analysis is slow and expensive. After several days, the Swiss chemical companies found large quantities of dioxin in the samples from the plant. What was less clear was how far the dioxin had spread, and at what levels. During that first week, ICMESA workers were directed to take samples at distances farther and farther from the plant. The analysis of these samples indicated that levels of dioxin remained high at greater distances from the plant, but there was little consistency to the pattern of deposition.

Givaudan also contacted experts from firms familiar with dioxin incidents in Britain, West Germany, the Netherlands, Austria, and the United

States. Those companies consistently advised Givaudan to evacuate the residents as soon as possible. Apparently deciding that the geographical limits and specific areas of high dioxin contamination remained too uncertain, and fearing panic among the people, the Swiss chemical companies held back on sharing the sample results or on recommending an evacuation to the Italian authorities. The decision was soon to be taken out their hands.

The following Saturday, Dr. Aldo Cavallaro, a health official from Milan, visited the plant. ICMESA representatives explained that the explosion occurred as a result of a temperature increase in the TCP reactor vessel. Returning to his lab, Dr. Cavallaro reviewed technical literature on TCP production and discovered that elevated temperatures can lead to the formation of dioxin, known to be an extremely toxic substance. On Monday morning Dr. Cavallaro spoke with ICMESA representatives and asked if dioxin was present in the cloud. When informed that it might have been, Dr. Cavallaro flew that same day to Switzerland to further question Givaudan management about the possible presence of dioxin in the fallout. He was told that the samples did in fact contain dioxin, but that the levels were unclear.

Armed guards and barbed wire kept people out of the restricted zones in Seveso.

Credit: ©AP-Photo/EM/STF/FM

Also on Saturday, the media picked up the story. Givaudan, and its parent, Hoffmann-La Roche, were still equivocating on the levels of dioxin. In the meantime, the media checked with technical experts, who reported that as little as 5 pounds, the amount believed at that time to have been released at Seveso, was enough kill more than 100,000 people. Such reports naturally created panic in Seveso.

As the health and environmental effects became clearer and more widespread and media coverage grew, action was finally taken. Officials required local doctors to report any illnesses in the area, and consumption of food from the area was forbidden. Some health monitoring was scheduled, and national health officials provided information on dioxin to local authorities and sent technical support. The authorities also conducted further testing of vegetation to determine the levels of dioxin. On Wednesday, July 21, ten days after the explosion, the ICMESA managing director and Paolo Paoletti, the production director, were arrested and criminally charged with causing the disaster.

Despite the increasing intervention through the second week after the explosion, health authorities continued to issue reassuring claims that everything was under control and that no drastic measures were necessary. When a spokesman for Hoffmann-La Roche announced that the situation was very serious and that extreme measures were necessary, including the evacuation of people and the destruction of homes, a local official was dismissive, declaring that "this man has been parachuted in; nobody was expecting him . . . I have the impression he is bluffing."[2]

Meanwhile, birds fell dead from the sky, pets seemed to walk drunkenly, and more animals died. Plants turned brown, as if burned. Rabbits oozed blood from their mouths and rectums. More children developed sores on their bodies, with more than a dozen hospitalized, and adults complained of nausea, vomiting, liver and kidney pains, and acne.

By the second Saturday, officials determined that an evacuation had to be ordered and that further health studies were necessary. With the dioxin deposited unevenly, the difficulty was in determining where to draw the boundary for the areas to be evacuated. Compounding these difficulties was the immense task of setting up and carrying out medical testing and a health-monitoring program for perhaps thousands or even hundreds of thousands of people.

The authorities divided the town into zones, according to varying degrees of toxicity. The most serious was Zone A, immediately south of the ICMESA plant, where evacuees were directed to take only their most essential possessions and to leave behind all household materials, food, pets, and livestock. In Zone B, which contained five thousand people, adult residents were informed that they could remain but that all children under fifteen and all pregnant women were to be removed from the area during the day and returned at night, to reduce their exposure to the dioxin. They were advised not to eat any fruit, vegetables, or animals and not to touch the ground.

Finally, the authorities informed residents that the cloud and its fallout crystals had contained something called dioxin, which they were now learning was among the most dangerous substances known. They quickly figured out for themselves that this dioxin had been present for over two weeks in the dust, on the ground, in the grass, on the fruit and vegetables, and even perhaps in the meat of their rabbits, chickens, ducks, and goats.

Italian Army soldiers strung barbed-wire barriers around Zone A. A fence was installed along the autostrada, the superhighway that connects Milan with the resort area of Lake Como and runs right through part of the contaminated area. Soldiers patrolled the autostrada. Signs were posted to warn drivers: "CONTAMINATED AREA. ROLL UP WINDOWS. CLOSE VENTS. DO NOT STOP. DRIVE SLOW." A sign pointing to Seveso was covered over with a skull and crossbones. The only observable presence in Zone A was the workers in white decontamination suits who were taking samples and collecting dead animals.

Residents were not told how long they would have to stay away from their homes in Zone A, nor for how long the children and pregnant women in Zone B would be taken away during the day. More than thirty people were hospitalized, mainly with skin lesions. Most disturbing was the uncertainty about their future health, especially the health of their children—both those already born and those yet to be born.

The population was evacuated from Zone A between July 26 and August 2 and housed in hotels outside Milan, with whole families crammed into small rooms. They were treated as pariahs, consigned to corners of restaurants and served with gloves. The psychological strain was intense. Uncertainty as to when, or even if, they might be allowed to return home contributed to the stress of the situation. The people lived like this for a year.

While the refugees endured such conditions, Givaudan embarked on a public relations campaign that involved keeping the company quiet and shifting the burden of dealing with the residents and the contamination to the public authorities.

The fallout from the dioxin cloud had been erratic, subject to shifting winds and downdrafts, so that there was no uniform distribution of the dioxin. As a consequence, more and more soil samples revealed additional hot spots of dioxin. Zone A was extended to most of Seveso. At first, about 225 people were evacuated, but with the expansion, another 500 were moved. Zone B was also expanded, and authorities created a third area, consisting of several thousand acres and 20,000 people who were to be monitored, but not restricted, in their movements. This area was designated as Zone R, for Zone of Respect. With each enlargement necessitated by more sampling, the people in the area adjacent to the identified zones wondered how long it would be before they also became zoned.

Zone A became a wasteland, surrounded by miles of barbed wire and guarded by armed soldiers. Over 2,000 rabbits, as well as other livestock and pets, died within a short time. The animals that did not die were slaughtered by the authorities—over 50,000 of them. Hunting throughout the area remained prohibited for eight years after the explosion. Even beehives were destroyed.

Once the people were moved out of Zone A, officials moved to assess the health effects and future risks, and to develop a plan to clean up the contaminated areas. Indoor air sampling of area schools was taken each month for one year, while soil monitoring was carried out for ten years after the explosion. Soil monitoring was designed to measure accurately the distribution of dioxin in the area, to provide background data for risk assessment, and to determine if cleanup measures were effective. In addition, epidemiological information was collected—informally and incompletely at first, and then more systematically. All of this was enormously difficult and expensive, and equipment was in short supply. For instance, mass spectrometers were critical to dioxin analysis, and there were only five such machines in Italy.

The most obvious health effect of the dioxin was the burning of children's skin, resulting in rashes and swollen faces. These skin conditions showed up within several days of the incident and shortly became widespread, affecting hundreds of children and requiring the hospitalization of dozens. At first, the skin lesions seemed to clear up within several weeks. However, by late August, chloracne began to appear. Chloracne is an acne-like skin condition in which the skin develops blackheads, papules, and pustules, mainly on hairy parts of the body.

Almost 200 cases of chloracne, particularly among children, were reported, accompanied by gastrointestinal illnesses, including nausea,

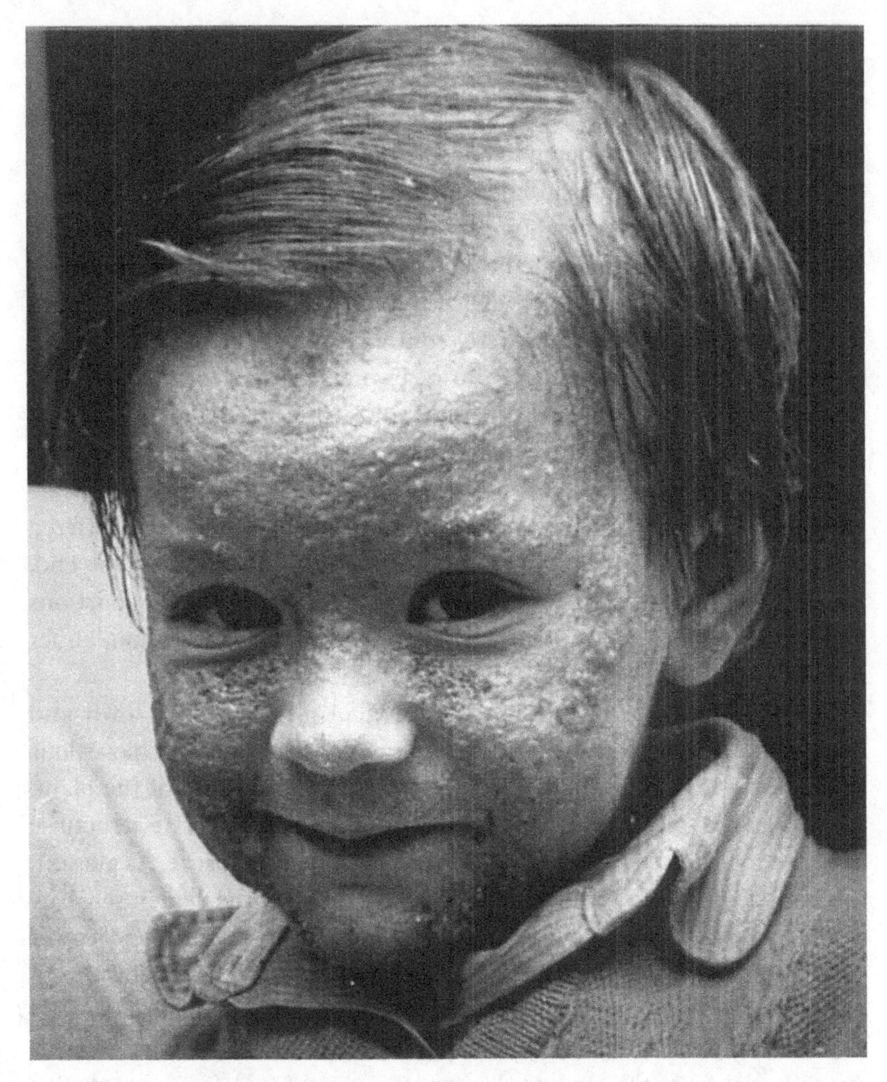

This child and many others in Seveso were afflicted with the skin condition chloracne as a result of exposure to dioxin.

Credit: ©AP-Photo/WM/STF/Fornezza

vomiting, abdominal pain, and gastritis, as well as headaches and eye irritations. Officials also discovered a higher incidence of chloracne in parts of Zone R, which people believed to be the least contaminated of all the zones. It was becoming apparent that the boundaries that had been drawn between the more- and less-contaminated areas were arbitrary.

The impact of dioxin did not stop with chloracne. Widespread reports (based on animal studies) that dioxin caused grotesque deformities in newborns conjured up images of thalidomide babies. Indeed, autopsies of chickens and poultry from Seveso revealed pathological conditions that had never previously been seen. Local doctors warned against conception and advised pregnant women to consider having abortions.

Abortions were illegal in Italy at the time, but a recent court decision had cleared the way for abortions when the mother's health was in danger. By mid-August, a regional medical commission recommended voluntary therapeutic abortions for those exposed to dioxin within the first trimester. Even in such cases, women were required to appear before a committee of two doctors and a psychiatrist to receive permission for the abortion. The civil authorities were not the only concern. The Catholic Church, a powerful influence in Italian society, opposed all abortions, even for the Seveso women. The Archbishop of Milan reportedly solicited volunteers to adopt any dioxin "monsters" that might be born. Other clerics offered consolation to those pregnant women who chose abortions. In the end, about thirty-four women received permission for therapeutic abortions and perhaps another hundred had abortions without permission, either illegally in Italy or elsewhere.

One month after the explosion, the ICMESA plant was shut down, and the question facing the authorities was how to dispose of the hazardous materials. It took almost six years to complete the cleanup of the plant. Initial plans called for encasing the entire plant in a concrete structure, but the people in the region rejected any permanent enclosure. As a result, the company and the authorities decided to dismantle the plant just as one would a nuclear power facility: seal the windows, doors, and cracks; lower the air pressure inside the building; and send in workers in airtight suits to dismantle the reactor and pack the dioxin-contaminated material in lead drums. This nuclear method was implemented in 1982. The most hazardous material from the reactor cleanup was packed into forty-one drums to be disposed of elsewhere. These forty-one Seveso drums were later to become infamous throughout Europe.

Authorities also had to decide when to let residents return. Some experts wanted to write off the entire contaminated area, but those who had built their homes and lives in Zone A wanted them back. The residents wanted the area restored to the conditions that had existed before the dioxin explosion. It was agreed that the worst-hit areas of Zone A had to be leveled and the least-contaminated areas, where about 60 percent of the displaced people had lived, would be restored. Anything replaceable was thrown away: curtains, carpets, clothing, furniture with upholstery, wooden floors, food, and appliances. Walls were repainted, floors refinished. Building exteriors were washed and the contaminated soil was removed. By 1978, residents from some parts of Zone A were permitted to return to their homes, although the homes were hardly the familiar ones they had left more than a year before.

The remediation in Zones B and R began in 1977. Soil was either removed or was plowed under so that cleaner soil deeper down would mix with the dioxin-contaminated soil, thereby lowering the concentration of dioxin. Fresh, clean soil covered the affected areas.

Once the inhabited zones were rehabilitated, the more contaminated areas of Zone A were addressed in 1982 and 1983. Buildings were torn down and vegetation and topsoil were removed. All of this material was disposed of in special concrete basins. Land that could not be cleaned, or diluted, to below five micrograms per square meter was fenced off. Eventually, over 270,000 cubic yards of contaminated soil was removed and disposed of in the concrete basins.

The emergency conditions throughout the Seveso area put a severe economic strain on the resources of the local and national governments and virtually destroyed the local economy, which was dependent on crafts and agriculture. The community directed much of its frustration and anger at the chemical companies and at the local officials. Shortly after the incident, the managing director, the company chairman, and three others from ICMESA and Givaudan were criminally charged with negligence, causing contamination, and a failure to have safety systems. All five were convicted by an Italian court, and each was sentenced to several years in prison. However, three of those convicted won reversals on appeal, and the others had their sentences suspended.

On several occasions, there were also violent responses to the events at Seveso. In May 1977, a local health officer was shot and wounded in the legs when intruders burst into his office seeking records on ICMESA and other companies. Two months later, on the first anniversary of the explosion, terrorists bombed the home in Switzerland of a Hoffmann-La Roche executive who had been responsible for dealing with the regional Italian authorities with regard to Seveso. The most serious act of retaliation took place in February 1980, when the 39-year-old director of production for ICMESA, Paolo Paoletti, was murdered by Prima Linea (Front Line), an Italian terrorist group.

Anger and violent impulses surfaced again in 1982, when the forty-one drums of dioxin material were lost. The blue drums contained 2.2 tons of heavily contaminated dioxin material from the ICMESA reactor. Hoffmann-La Roche subcontracted the disposal of the drums to a French waste consultant, Bernard Paringaux. Paringaux picked up the drums in September 1982 and was escorted to the French border by Italian authorities. At the border, the Italians washed their hands of any responsibility for the drums. Paringaux crossed into France and represented to the French custom officials that he was carrying a load of "halogenated aromatic carbons." He did not mention Seveso or the dioxin. Paringaux told Hoffmann-La Roche that the wastes were properly disposed of, but he would not tell them where or how.

Later in 1982, a Greenpeace activist heard that the drums were going to be dumped into the Atlantic Ocean. When Greenpeace made demands as to the whereabouts of the drums, they discovered that no one knew where they were. The Italians said that the drums were not in Italy; Hoffmann-La Roche said that they were properly disposed of—somewhere. In March 1983, a French magazine claimed that the Seveso drums of dioxin had been brought into France. Paringaux refused to talk and was jailed.

Accusations, threats, rumors, and investigators all flew across the borders between Italy, Belgium, France, Germany, and Switzerland. After seven weeks in prison, Parginaux revealed, in June 1983, that he had stored the forty-one dioxin drums at an abandoned slaughterhouse outside Paris. French health authorities seized the drums and moved them to an army camp. Hoffmann-La Roche had already been the target of boycotts and protests outside their plants, and several of its employees had been attacked and one killed. Their negligence with the forty-one drums prompted more anger, which did not help their sense of safety, or their reputation. Not surprisingly, Hoffmann-La Roche agreed to have the drums shipped to Switzerland, where they were incinerated at a disposal facility owned by Ciba-Geigy, another Swiss chemical/pharmaceutical company.

As a result of the fiasco—an embarrassment not only to Hoffmann-La Roche, but also to the Italian and French governments—the European Common Market (now the European Union) passed the Seveso Directive, which required chemical companies to provide full information on hazardous sites and the storage of dangerous substances, and on potential risks from the sites and how those risks could be reduced. In addition, several years later, Italy enacted legislation that reformed its national health care system to include standards for the production, registration, sale, and use of chemicals that affect biological and ecological systems; a national inventory of chemicals; and risk maps that require factories to provide

Drums of dioxin-contaminated materials from the cleanup in Seveso were discovered in an abandoned slaughterhouse outside Paris.

Credit: Dino Fracchia, courtesy of the artist

toxicological data on their products, and possible impacts on people and the environment.

The Swiss companies early on accepted responsibility for paying the costs of housing the refugees from Zone A. In addition, over 7,000 private lawsuits by individuals were settled out of court. Including the cleanup costs, payments to various government authorities, and settlement of the private lawsuits, Hoffmann-La Roche spent more than $162 million. Yet, in 1982, when Givaudan was paying out these settlements, its parent, Hoffmann-La Roche had sales of $3.6 billion.

By the late 1980s, the restrictions that applied to Zone R were lifted, Zone B was cleaned up, and a part of Zone A was turned into a park. Financial settlements between the governments and the chemical companies were resolved, and the criminal proceedings concluded. After the first ten years, from 1976 to 1986, there was no evidence of a dramatic increase in birth deformities attributable to dioxin exposure. The chloracne attacks receded, although scaring remained visible on 15 of the more than 150 originally afflicted. As yet, no clear, significant increase in the incidence of cancer or other diseases attributable to the dioxin was detected. Indeed, an international committee, which had been established to monitor the effects, ended its work in 1984, concluding that chloracne was the only health effect resulting from the dioxin exposure at Seveso. There were, though, anecdotal reports of an increase in deaths from cardiovascular diseases among the affected population, and epidemiological studies had not yet reported results.

Dr. Pier Alberto Bertazzi and colleagues at the Institute of Occupational Health, Epidemiology Section at the University of Milan conducted epidemiological studies at Seveso. They selected as a cohort study group more than 30,000 people, with about 700 from Zone A, 4,000 from Zone B, and 26,000 from Zone R. As a comparison or control group, 180,000 people from the wider area who had not been exposed to the dioxin were selected. Researchers collected vital statistics, medical records regarding any occurrence of cancer and other diseases and, eventually, cause-of-death data for each member of the cohort and control groups.

In 1989, Dr. Bertazzi and his colleagues reported significant increases in mortality from heart disease in males, with the highest increase in Zone A, which confirmed the anecdotal evidence from local doctors. It was hypothesized that both the dioxin exposure and the stress from the disaster contributed to this increase. Bertazzi also found higher than expected incidences of liver cancer, tumors relating to the blood, and soft-tissue sarcomas, especially among those residing for the longest period in the contaminated area.

Based on further studies of the cohort, in 1993 Bertazzi reported elevated risks for several somewhat rare cancers attributable to the Seveso dioxin. Women in the Zone B population were subjected to a five-fold increase in the rate of gall bladder cancer and multiple myeloma, a rare bone-marrow cancer, and men were subjected to the same rare cancers at a two-fold increase. In Zone R, Bertazzi found elevated incidence of soft-tissue sarcoma and of a rare non-Hodgkin's lymphoma.

In follow-up studies that were conducted ten and fifteen years after the exposure, no increased risk of breast cancer was found. However, a more recent study found a statistically significant risk for breast cancer related to the concentration of dioxin in the blood of women who lived near the site of the exposure in 1976. Blood levels of dioxin typically are found at several parts per trillion. The Seveso women with breast cancer had concentrations of 13 to 1,960 parts per trillion in 1976.

Despite the findings about dioxin's carcinogenic effects, its real danger may lie in its reproductive, developmental, and immunological impact.

Based on animal studies, it has been hypothesized that dioxin actually disrupts multiple endocrine systems, affects cell growth, disrupts fetal development, and suppresses immune systems. Studies have shown that dioxin, and related chemical compounds, can disrupt hormones and growth factors, decrease fertility, affect central nervous systems, and even alter certain learning behaviors. And since dioxins seem to affect cell growth and disrupt fetal development, the *in utero* effects of exposure to dioxin are especially disturbing. Children of women exposed have shown a variety of developmental effects (smaller size and abnormalities of gums, nails, skin, teeth, and lungs) and delays in psychomotor development. A distorted ratio between births of males and females for the residents of Seveso has also been reported. Between 1977 and 1984, only 35 percent of children born to the most exposed adults were boys, and no boys were born to the parents with the highest levels of contamination. This decline in male births at Seveso is cited as a very specific instance of the mysterious decline of male births throughout the world since 1970.

Some twenty-five years after the environmental disaster in Seveso, the people exposed to dioxin continue to be subjected to disturbing health effects. The uncertainty that settled on their lives, like the cloud from the factory, remains with them.

LOVE CANAL, NEW YORK 1978

In the late 1970s and early 1980s, an area called Love Canal in Niagara Falls, New York, became America's most infamous toxic waste site. Media coverage at the time showed images of holes in backyards filling with thick, black, substances; toxic chemicals entering basements through sump pumps and walls; a grade school closing because of the danger to children; angry citizens screaming at local, state, and federal officials to do something; housewives taking officials hostage. Over 230 families living next to Love Canal were evacuated in August 1978 because of the health risks associated with the over twenty thousand tons of toxic chemical waste that had been dumped in the canal by a chemical company in the 1940s and 1950s. By 1980, when the dangers of the chemicals were better understood, the evacuation was expanded to cover an even wider area.

The canal's beginning was less notorious. It was dug in the 1890s by William Love as part of a proposed power scheme in the Niagara Falls area, but the project failed when it was only partially completed. Other power projects did succeed in harnessing the water from the Niagara River, bringing cheap hydroelectric power to the area. This, combined with a large supply of salt, attracted the Hooker Electrochemical Company in 1906. Hooker manufactured chlorine and caustic soda, used for bleaching, disinfectants, paper, and soaps, but the company did not make money for the first several years.

World War I changed Hooker's prospects. Germany had monopolized the chemical industry, and when the war cut off supplies from Europe, Hooker and other American electrochemical plants leaped into the breach. By the end of the war, Hooker was producing seventeen chemicals and manufacturing synthetic dyes, perfumes, and medications from coal tars. Net profits in 1918 were $1.34 million.

World War II boosted Hooker's fortunes, just as the First World War had. Hooker supplied chemicals to make smoke pots, colored flares, disinfectants, military shoes, and lubricating oils to keep the machines of war running. After the Japanese captured 90 percent of the world's natural rubber supply, Hooker supplied dodecyl mercaptan to the government for the production of synthetic rubber. Thionyl chloride and arsenic trichloride produced poison gases. Hooker was perhaps most proud of, and secretive about, the chemicals the company manufactured for the Manhattan Project, which were used for making the atomic bomb.

The expansion of business increased waste residues from the chemical processes that had to be disposed of somewhere. By the early 1940s, when Hooker had little room left on its own plant property, it found Love Canal.

The canal was fed by an artesian spring. Watercress, boysenberries, and apple and cherry trees grew along the property. Homes were built in the area, and in the summer, girls and boys swam in the canal. In the winter, residents ice skated on the canal's frozen surface. The canal stretched three thousand feet south to north, was about sixty feet wide, and was ten feet deep.

Hooker acquired the rights from successors to Love's company to use the canal and started dumping in 1942 in the northern section, between what is now Read Avenue and Colvin Boulevard. Fifty-five-gallon drums were filled with solid and liquid residues at the Hooker plant, loaded onto trucks, and dumped into the canal. Hooker constructed dams along a portion of the canal that was used for dumping, sometimes pumping water out of the dammed-off section in order to dump in drums of the chemicals, and other times emptying the drums of chemicals directly into the water. Hooker also dug pits adjacent to the southern section of the canal for dumping chemicals. Some of these pits were dug within several feet of residential backyards.

The drums, usually old and rusted, were dumped randomly, often breaking open and spilling their contents. The residues filled the pits and portions of the canal nearly to the level of the original ground surface; afterward Hooker would place dirt, and occasionally ash, on top. The effect of all this was to create conditions under which the ground slowly caved in. With drums lying every which way, with spilled liquid wastes mixing with ash and clay and dirt, and with old, rusted drums deteriorating, breaking, and spilling more chemical contents, the ground subsided and potholes appeared, and the dangerous contents of the drums rose to the surface.

Over the years, Hooker management had gained extensive, specific, knowledge of the dangers associated with its chemicals and their residues. By the 1930s, arsenic trichloride was known by Hooker management to be so poisonous that exposure could result in vomiting, inflammation of the skin, loss of hair, and liver and kidney problems. By the 1940s, thionyl chloride was known to be highly reactive; upon contact with the air it created a fume of hydrochloric acid and sulfur dioxide that burned people. Both of these chemicals were used to make war gases and both were dumped at Love Canal. So, too, was mercaptan—the chemical that was used to produce synthetic rubber—which caused nausea, vomiting, diarrhea, and blood in the urine. In all, Hooker dumped more than 200 chemicals at Love Canal.

Within two years of the start of the dumping, chemicals began to surface. Hooker's Annual Operations Report for 1944 stated that burying its residues was "creating a potential future hazard" and predicted that "eventually we will have a quagmire at the Luve [*sic*] canal which will be a potential source of law suits in the future." In August 1946, several key managers from Hooker inspected Love Canal and reported to the president of the company that the entire length of Love Canal was filled with water that appeared to be contaminated, and that children in the neighborhood used the water for swimming. The managers advised the company to fence the property and put up warning signs. Hooker did neither.

During the dumping, residents witnessed fires that shot as high as the houses next to the canal. Explosions at the dump sent burning material up to two blocks away. The proximity of the dumping to homes meant that horrible-smelling, rainbow-colored liquids ran off the canal property and into backyards. Dust, white powder, and ash blew from the dump onto homes. The odors were so foul and pervasive that it led to another inspection by Hooker.

In October 1950, a representative of Hooker reported to management that the ash being dumped at Love Canal was blowing toward houses east of the canal. He observed that the water in the canal was contaminated by an "oil slick and large globules of congealed residue covering most of the surface of the pond," and that "the ground had settled enough to open pot holes of various depths and that portions of buried drums were exposed in these holes." Potholes and exposed drums were found at spots that had been filled and covered as little as a few years previously. The representative reported to top management that "it is felt that a fence around this property would be very desirable from a safety standpoint." No fence was put up.

By the spring of 1952, Hooker knew that the drums in which the chemical residues were buried were in poor shape, and would continue to deteriorate; that the water in the canal was contaminated; that potholes or sink holes had appeared, exposing the drums; and that with time chemicals would rise to the ground surface.

In the spring of 1952 the Niagara Falls school board asked if Hooker would consider selling a part of the Love Canal property. The baby boom had reached the area, and both homes and a new grade school were needed. Hooker initially rejected the idea after top management was advised that the company should look for another dump site and discontinue using Love Canal, that plans should be made to prevent the property from becoming a nuisance, and that it was too risky to sell Love Canal.

Many of the operations staff were distressed at the idea of building a school on a toxic waste site. They knew that chemical wastes were dangerous if disturbed and that subsidence would continue to occur for a long time, so that the wastes would become dangerous even if they were left alone. And they knew that the school board was in no position to manage such a place. The plant superintendent at the time, who later became president and chairman of the board of Hooker, stated:

> [T]here was a general knowledge that these organic chemical residues that we were disposing of was a mixture of all kinds of things, who knows what, and it was in the ground all mixed together and we just had a general feeling that, by golly, it better stay there and we better keep control of it to be sure it stayed there. That was just a general feeling that we all had.[1]

Yet less than a month later, Hooker decided to transfer the property to the school board. As one of the managers responsible for the decision wrote at the time:

> The more we thought about it, the more interested... [we] became in the proposition and finally came to the conclusion that the Love canal property

> is rapidly becoming a liability... [we] became convinced that it would be a wise move to turn the property over to the schools provided we would not be held responsible for future claims or damages resulting from underground storage of chemicals.

Thomas Willers, the comptroller of the company in 1952, who attended the meetings of the Management Committee that was responsible for dealing with the school board on Love Canal, later provided an explanation for why Hooker reversed its decision. Willers described how, after the initial contact and Hooker's refusal, the company looked more closely at its requirements for waste disposal and decided that they could manage without Love Canal. It was close to maximum capacity anyway, alternative sites were available, and Hooker could insist in its agreement with the school board that it be able to continue using Love Canal for a while. Willers further recalled, "I don't think the property was all that valuable anyway... I'm talking dollars and cents...." Hooker recognized that the area was rapidly developing, and using Love Canal as a dump was becoming a liability. So Hooker gave Love Canal to the school board in return for a provision in the deed that protected the company from any liability there.

One major obstacle, however, was the vociferous opposition to the sale among the plant managers. Because the managers were valued employees, Hooker told them that the transfer of Love Canal had been forced upon the company by the school board, which was going to condemn the property if Hooker did not sell it. One manager was told:

> Since the school board was going to take it anyway, we would be smarter to give it to them, in return for which we could get strong statement which would protect Hooker from damage suits if something happened after the school board acquired it.

This explanation, however, did not accurately reflect the negotiations with the school board. It is likely that Hooker's top management simply concocted the story about a condemnation threat to rationalize to its own people a decision that was laden with problems.

Even though the school board only requested a part of the site, in April 1953 Hooker transferred the entire property to it. Hooker advised the board that the unfilled central section was suitable for installing foundations for a school. Hooker, however, made use of its right to continue to use Love Canal for dumping waste materials until February 1954, including

In this Department of Health aerial photograph, the bare, unvegetated areas directly behind the homes are where chemicals surfaced. The 99th Street School is the large building on the right, in the middle of the photo, with ball fields behind it.

Credit: Courtesy of the New York State Department of Health

part of the central section. The City of Niagara Falls also dumped municipal waste into the canal during this period.

Almost immediately after Hooker transferred Love Canal to the school board, the consequences were felt like an aftershock. In January 1954, when Hooker was still dumping at Love Canal, a contractor, excavating the foundation for the new grade school on 99th Street, encountered a pit filled with black water. This was in the central section, the very area in which Hooker had advised the school board that it would be safe to build. As a result of the chemical wastes, the school was moved about eighty-five feet north, but eventually swings and other play equipment were installed on top of the area where the chemicals had been found.

In May 1955, after the 99th Street School had opened, about twenty-five square feet of ground crumbled near the original excavation, exposing drums and chemicals. Some of the children were splashed and their eyes were burned. The school principal called Hooker to ask for information about the chemicals, and a Hooker representative was sent to investigate, along with the Hooker plant nurse. The nurse provided advice on appropriate first aid, and Hooker arranged for ten trucks of dirt and a bulldozer to cover and grade the exposed area.

In November 1957, the school board considered selling part of the Love Canal property to developers for the construction of homes. At the school board meeting Hooker opposed the idea because of concerns that developers would expose chemical waste. What Hooker did not mention, however, was that the chemical wastes had already begun to surface and constituted a more serious problem than the potential risks of new development. Hooker also did not mention that the subsidence problems would continue for decades, resulting in further exposure of toxic chemicals. And though it had recently received disturbing news about one of the chemicals contained in the site, Hooker divulged nothing about the nature of the waste.

As early as the 1940s Hooker workers had experienced outbreaks of dermatitis and chloracne as a result of their exposure to chlorobenzenes, arsenic trichloride, and, especially trichlorophenol (TCP), the same chemical that caused the disaster in Seveso, Italy. Chloracne is a skin condition that produces extremely disfiguring pimples, boils, or pustules that recur and can be very painful. They develop around the eyes and ears, but also on the back and chest and even in the groin area, and can continue for years, even decades.

In the mid-1950s, Hooker was contacted by a customer who had purchased its chemicals for use in a weed-killer product. The customer reported incidences of chloracne in its manufacturing facility and among some people who were using the weed killer. The customer asked for confirmation that the chloracne was likely caused by an impurity in the trichlorophenol process. Hooker replied that it believed the impurity occurred as a result of high-temperature boiling in the TCP process.

In April 1957, the director of a German company delivered some disturbing news. The company had been conducting extensive studies in conjunction with a hospital in Hamburg and had traced the impurity in the TCP process, which was causing the chloracne, to a chemical reaction that led to the formation of dibenzodioxine. We know this compound as dioxin. The Germans also reported to Hooker that the dioxin was extremely poisonous, and that all possible precautions to prevent exposure to it should be taken. Where major spillage had occurred, several companies had to decontaminate entire buildings by removing all the insulation, chipping off old paint, and tearing up and replacing floors. The representatives of the German company described dioxin as having "a really sinister character."

Although Hooker had dumped over 250 tons of trichlorophenol, containing dioxin, at Love Canal over the years, the company did not pass this information along to the school board. At a meeting in November 1957, the school board decided to not pursue the plan to sell off part of the property at that time.

Less than a year after the school board meeting, children again suffered burns from chemical exposure. Hooker investigated and determined that

Children discovered "elephant's footprints" where the ground subsided and colored, toxic chemicals seeped to the surface.

Credit: Courtesy of the New York State Department of Health

the ground had subsided, exposing drums, and leaving benzene hexachloride (BHC) on the surface. They also saw that the entire Love Canal property was being used by the children as a playground. Children had even picked up the chalk-like BHC cake and had rubbed it in their eyes, which had burned them. Chemicals had also surfaced at homes adjacent to the canal.

Aileen and Edwin Voorhees had lived on 99th Street, adjacent to the canal, since the early 1940s. In 1958 they built a new home and soon began a difficult and ongoing struggle with toxic waste. No matter what they tried, the Voorhees could not stop "thick, black, smelly stuff" from seeping into the basement of their house. Waterproofing the walls did not work, neither did digging a trench around the inside of the basement walls and draining the chemicals. As Edwin Voorhees later described, "all of a sudden you get these chemicals coming through... in the northeast corner of the house... and you also had them coming in the other side, so the only alternative I thought I had was to put another sump pump in and try to pump them away." But they would not go away.

Meanwhile, back on the Love Canal property, a drum of thionyl chloride had exploded, spewing chemicals, and drums of BHC had surfaced as a result of more subsidence. Children continued to play on the site, throwing the tops of drums like disks. They threw lumps of white powder, which burned them, and chunks of material they called "fire rocks" that sparked or exploded when thrown against other objects.

In the late 1960s, the northern section of Love Canal was transferred to the City of Niagara Falls for recreational purposes, and the southern section was sold to a private individual who never developed the property. No homes were built directly on top of the dump, only on land directly adjacent to the Love Canal property. In 1968, the State of New York acquired a thin strip of land at the very southern tip of the canal as part of an expressway construction project. During construction, the state encountered contaminated soil and chemical waste that it removed from the site.

Chemicals not only rose to the surface as a result of subsidence, but also moved through the ground. The stiff clay soil at Love Canal, extending from about five to twelve feet below the surface, was fractured, providing an easy pathway for chemicals to move away from the dumping site and into the adjacent properties. Residents started to encounter black, chemical water when they dug postholes for fences. Karen Schroeder, the daughter of Aileen and Edwin Voorhees, moved into a house just up the street from her parents. In October 1974 the built-in fiberglass swimming pool in the Schroeders' backyard suddenly rose several feet out of the ground. When the pool was removed, the hole filled in with chemicals.

Peter Bulka, a local policeman, lived with his family in a home adjacent to the canal. In 1969, Bulka was managing a Little League baseball team that played on a diamond located on the northern section of the Love Canal property, near Colvin Boulevard. During a practice, one of the players came running in after chasing a ball and shouted something about volcanoes. Bulka and the others went out to look and saw that little volcanoes, spewing a light gray fume that smelled like thionyl chloride, had appeared in the outfield. The baseball field was subsequently moved away from Love Canal.

Bulka's own two-year-old son, Joey, fell headfirst into a pothole on the canal property, and might have drowned if an older brother had not pulled him out. Joey's face and neck were covered with a black soot-like

A resident pours black sludge taken from the ground near the 99th Street grade school.
Credit: Courtesy University Archives, State University of New York at Buffalo

substance that smelled of chemicals; he later developed an ear impairment. In the mid-1970s Bulka was forced to replace the sump pump in his house three times before finally having one specially constructed to withstand the attack of the chemicals. Nonetheless, two of his children developed severe reactions to the chemicals in the basement. Bulka was experiencing firsthand how Love Canal was quickly becoming the quagmire that some Hooker employees had predicted in 1946, and an entire neighborhood was about to disappear into that quagmire.

Bulka was not given to complaining, but he finally took his concerns to the local health authorities in the summer of 1976. His neighbors were not pleased. They worried that if the authorities decided to pursue his complaints, it would negatively affect their property values. Moreover, many of the neighbors worked at chemical companies in Niagara Falls, which were a major source of tax revenue for the city. A complaint about one of the companies was seen as a threat to the livelihood of the community.

In the summer of 1976 New York State's Department of Environmental Conservation (DEC) investigated the contamination of fish in Lake Ontario from the chemical mirex. As a part of the investigation, the DEC attempted to identify any companies whose present or past disposal practices might have contributed to the contamination of the Niagara River, which feeds into Lake Ontario. Hooker was identified as the only manufacturer of mirex in the area.

DEC staff visited Hooker to ask about any current or former disposal sites that might be discharging contaminants into the river. Hooker officials mentioned several dumping sites, including Love Canal, but reiterated "that they have no legal responsibility for Love Canal." Arrangements were made for the DEC to visit the Hooker plant facilities disposal sites, including Love Canal, and to take samples. The DEC also requested that Hooker provide information on the identification, volume, and location of the chemical wastes dumped at Love Canal.

In October 1976 the local newspaper, the *Niagara Gazette*, began an important series of articles covering the "industrial horror story" at Love Canal. They reported on the black, oily substances that were ruining Peter Bulka's sump pumps, even hiring a consultant to test the sump. That November, the *Niagara Gazette* reported that the results from those tests revealed that fifteen organic chemicals were found in the sample from Bulka's basement sump, including three toxic chlorinated hydrocarbons. In the community adjacent to Love Canal, sumps discharged into sewers that emptied into the Niagara River. One county official was quoted as saying that since the toxic materials were entering the sewer system from private homes, it was the responsibility of the homeowners to stop the discharges. This early affront was an inauspicious start for the emerging relationship between the residents and their governmental representatives.

The DEC took samples from the sumps of homes adjacent to the canal, from sewers in the area, and from the surface of the former canal. While sampling, DEC staff noted strong odors of chlorinated aromatic chemicals and a black, gummy sludge coating the sumps. When the samples were analyzed, they indicated the presence of significant quantities of a variety of chlorinated hydrocarbons. During this time, the city conducted a house-to-house survey of homes adjacent to the canal and found that the chemical invasion was pervasive. Discussions were initiated by the state with the City of Niagara Falls, the school board, and Hooker to develop a plan for dealing with the problem, and to determine who was going to pay for any cleanup.

In the fall of 1977 the state requested the assistance of the federal Environmental Protection Agency (EPA) to conduct further studies of the subsurface conditions and to monitor the air in the basements of homes adjacent to the canal. The EPA representative who inspected the site found conditions to be not only unhealthy and hazardous, but unprecedented in scope. He concluded that temporary measures would only delay resolving the problems at the site, and that, since it might be years before conditions were normalized, serious thought should be given to purchasing some or all of the homes.

Nature soon aggravated the hazardous conditions at Love Canal. In what became known as the blizzard of '77, the Niagara Falls area suffered one of the worst winters in its history. When the accumulated snow finally melted in the spring, it infiltrated the ground at Love Canal and exerted such pressure on the chemicals lying just below the surface that it accelerated their migration and surfacing.

Debbie Cerrillo, who had grown up in Niagara Falls, bought into one version of the American dream: a brand-new ranch home in her childhood neighborhood, with a large, open field at the back of her property, and a grade school on the other side. On occasion Debbie saw a ghostly green haze hanging over the open field, which was situated on the former canal, even though the rest of the area was clear. Although she thought it was strange, she paid no particular attention to it. At the beginning of the spring thaw in 1978, with snow still on the ground, Debbie went shopping on her snowmobile. On the way home, as she drove across the vacant field at the back of her property, the snowmobile suddenly got stuck in the field. Debbie got off and discovered that part of the snowmobile had sunk into a pool of black, horrid-smelling liquid that lay just below the surface of the snow. The substance got onto her gloves, and when she took off her gloves with her teeth, to try to free the snowmobile, some of it got into her mouth and made her gag. She recognized it as the same black material that had been reappearing for several years in the basement sump of her next-door neighbor, Peter Bulka. She also recognized the smell: when her father had worked at the Hooker Electrochemical Company in Niagara Falls, the smell would linger on his clothes when he came home after his shift.

That same spring, sampling results from the DEC and the EPA revealed disturbing levels of toxic chemicals on the surface and in the groundwater. The commissioner of the State Department of Health (DOH) inspected Love Canal and found the conditions on the surface of the canal to be deplorable: an acrid chemical smell was prevalent throughout; waste drums and their chemical contents were visible; and pools of black, tarry, oily liquids were found on the surface. Especially disturbing was the widespread presence of a toxic pesticide, lindane, on the surface of the canal property that was widely used as a playground by children. Following his visit, the commissioner issued an order to the local health authorities requiring immediate action.

The city found that it did not have the resources to cope with Love Canal. Hooker, meanwhile, kept a low profile, offering no comment on reports that the toxic chemicals at its former dumping site were entering homes. Internally, Hooker managers decided to "cooperate on any technical matters on which our advice is sought and provide general background information about the site, but avoid becoming actively involved in any remedial plans." The company indicated publicly that it would be more willing to cooperate if Hooker were insured against litigation.

The *Niagara Gazette* continued its coverage of events unfolding at Love Canal. The coverage eventually caught the attention of Lois Gibbs, a twenty-seven-year-old housewife who lived several blocks from the canal. Her son had recently developed unexplainable seizures after he began attending the 99th Street School. When she realized that the former dumping site being discussed in the paper was right next to her son's school, Gibbs became alarmed. The reports of dangerous chemicals that were buried next to the school and escaping into homes and the environment provided a possible explanation for her son's illness. Gibbs quickly requested a transfer for her son. With twisted bureaucratic logic, the superintendent told her that if her request were granted, it would be an acknowledgment to the wider public that the area was contaminated and would imply that all the children should be removed from the school. He did not transfer her son.

Gibbs, who had never actively organized anyone, started to organize everyone. She contacted a brother-in-law who was a biologist at the local state university and educated herself about the facts and issues, as well as they could be determined at that time. She knocked on doors in the neighborhood to see what others knew or suspected, and to find out what they intended to do. On one of her early rounds she met Debbie Cerrillo. After her exposure to the chemical problems at the site the preceding winter, Cerrillo was deeply concerned for her three young children, and she was not one to avoid a fight. Together they formed the Love Canal Homeowners Association.

In June 1978 government agencies held a public meeting to explain what was known about Love Canal. One local health official belittled the hazards from the landfill and told the residents that the devaluation of their homes because of the toxic chemicals was their own problem. Lois Gibbs attended that meeting and asked if the school was safe. When she got an evasive answer, she shot back, "Get this school down, it's contaminated!"[2]

By July 1978 the DOH was finally able to assess the health risks for people living adjacent to the canal. Those assessments indicated that the isolated risks from even a few of the chemicals were substantial and disturbing. The basements of homes registered high levels of chemicals as did the air around the dump site. Particularly noteworthy was the high level of lindane in areas used for play by children.

Another DOH study found a notably higher rate of miscarriage and congenital malformations among those living in the southern section, and an increased risk for spontaneous abortion among the women in both the northern and southern sections. It also found significant levels of toxic fumes in these same homes. The DOH assembled the available environmental and health risk data and submitted it to a panel of outside experts for an independent review. Events were now to take a dramatic turn.

Following that independent review, the commissioner of the DOH issued a public health order on August 2, 1978, at a press conference in Albany. The commissioner reported that more than eighty chemical compounds had been identified in various samples at Love Canal, including twenty-six organic compounds in air samples from the basements of homes.

Seven of the chemicals were carcinogenic in animals; one, benzene, was a known human carcinogen. Moreover, the epidemiologic study revealed an increased risk of spontaneous abortion among residents, especially among those living adjacent to the southern section, and congenital malformations among five children living adjacent to the canal. Based on these conditions, the commissioner recommended that all pregnant women and families with children under two years old living on the streets adjacent to the canal immediately relocate from their homes.

Many affected parties—government agencies, Hooker representatives, the press, and residents of the Love Canal neighborhood, including Lois Gibbs and Debbie Cerrillo—attended the press conference at which the commissioner announced his order. The residents thought they were attending a working meeting at which the problems at Love Canal would be discussed and that they would have an opportunity to participate in the resolutions of those problems, which were affecting them more than anyone else present. When they realized that the meeting had been called only for the purpose of delivering an order for a select few to evacuate, they were furious. After the commissioner read the order, Gibbs shouted out, "You're murdering us." When Cerrillo heard that only pregnant women and children under two were being recommended for relocation, she got up and demanded, "What about my two-and-a-half-year old; she's out of luck, right?"[3]

On August 3, 1978, the families subject to the DOH order started to relocate from their homes with the assistance of state agencies. On August 4, Gibbs, Cerrillo and others held a meeting to formally organize the Love Canal Homeowners Association.

New York Governor Hugh Carey contacted President Jimmy Carter and reported the events of the previous week, including the widespread risk to families in the area, and requested that the president declare an emergency. On August 7, 1978, President Carter declared a federal emergency, the first in United States history in response to an environmental condition. The federal and state governments also expanded the relocation to include the families on both sides of 97th and 99th Streets. Federal funds were committed to assisting the state in the relocation process, and the state government committed to permanently relocating all people living on both sides of the streets adjacent to the canal, and to buying their homes. Over 239 families were evacuated from Love Canal that year.

When a federal emergency was declared, some relief funds became available to begin the cleanup. In 1978 and 1979 a drain system was constructed around the canal to collect the leachate, contaminants mixed with groundwater, and divert it to a temporary treatment facility that removed the toxic properties and discharged the material to the city's sewer system. The site was also capped with clay to prevent more water from infiltrating the canal, thereby reducing the amount of leachate.

Further studies by the state, with assistance from the residents, indicated a high rate of both miscarriages and children with congenital defects among those living along the swales, or "wet" areas around the canal. As a result of the accumulated evidence, families with pregnant women or young children, and a low-income housing project west of the canal, were temporarily relocated to nearby motels at the state's expense.

The relocations were difficult. Families were consigned to motel rooms with two beds, a desk, a dresser, a TV, a bathroom, and cots for the kids, with little space for personal effects. Meals were eaten out. The motels were filled with families who shared the same anxiety about what awaited them when they returned home. By the end of 1979, the residents who had been temporarily relocated were allowed back in their homes.

Meanwhile, citizens continued to organize, publicize, and agitate for the permanent relocation of all those living near the canal. Cerrillo and Gibbs testified before Congress, while other residents appeared on the *Phil Donahue Show.* One resident attended Hooker's annual shareholder meeting and spoke out in protest. Cerrillo shook hands with President Carter during a reelection campaign stop at Buffalo airport, where the president told her, "I'll pray for you."

In the spring of 1980, events once again exploded at Love Canal. The EPA undertook a preliminary study of possible chromosomal damage for people living in the area of the canal. The study was conducted without any control group, and it was intended only to serve as a basis for deciding whether a full-scale, costly study was justified. These preliminary results, which indicated that eleven of thirty-six residents had chromosomal damage, inadvertently became known to the media, forcing the government to quickly inform the residents before they heard it in the press. No matter how much the government qualified the study as tentative and incomplete, the hard, cold number of eleven out of thirty-six overwhelmed the public. The study was later subject to widespread criticism among the scientific community, and the results were considered by many to be suspect.

When the EPA study was released, it understandably scared the remaining residents. The citizens had been arguing and pleading with the government to buy out their homes so they could get out. The chromosome study gave them leverage.

The Monday after the EPA study was released, residents met at the Homeowners Association office to follow developments. As the day passed, more and more people gathered. When the press reported that the White House was not going to relocate any more residents, the crowd grew angry and frustrated. Someone poured gasoline in the form of the letters "EPA" on the lawn of an abandoned house across from the association and set it ablaze. Others blocked traffic, which drew the police to the scene.

Gibbs called two EPA representatives, a doctor and a public relations man who were in the area, and asked them to talk to the residents about the chromosome study. When she announced that the EPA representatives were coming, someone in the crowd proposed that they take them hostage, to show them how trapped the residents felt. The EPA representatives arrived and tried to address the growing crowd, but there were more shouts to take them. Gibbs informed the doctor and the PR man that they were hostages of the "Love Canal People," and that they would not be harmed if they stayed inside. The hostages were fed homemade oatmeal cookies and sandwiches and were allowed to meet with individual members of the press inside the association offices. Gibbs called the White House and demanded the relocation of the residents before the release of the hostages. After about six hours of discussions with various officials, and with the FBI threatening to rush the crowd, the hostages were freed. Neither Gibbs nor anyone else was arrested for the action.

Within days, in May 1980, President Carter declared the second emergency at Love Canal, authorizing the federal and state governments to relocate about seven hundred families. Though only temporary relocation was specified in this emergency declaration, an agreement was reached in October 1980 to provide funds to buy more than five hundred homes in the area and to permanently relocate those who wished to move. Over four hundred residents accepted the offer; the others chose to remain.

After 1980, the state and federal governments focused on determining the impact of the chemical migration on the sewers and creeks in the area, and on how best to remedy the situation. They also undertook a study to determine whether Love Canal, after the cleanup, would again be a place where people could live and play. The study found that certain parts of the area could be habitable again, and it took over a decade to clean up the environment around the canal. Some of the homes that were evacuated in 1980 have since been sold and are once again occupied.

The United States and the State of New York sued Hooker to recover the costs of investigating and cleaning up the toxic waste site and of relocating and buying the homes of over five hundred residents. The litigation started in 1979 and went on for over fifteen years. During an eight-month trial in 1990–1991 many of the details of what had happened during the years of contamination were finally revealed. In the summer of 1994, the State of New York settled its claims against the chemical company in exchange for a payment of $98 million and an agreement by the company to assume the responsibilities for monitoring the site for as long as necessary. That settlement alone cost Hooker about $130 million. The chemical company and the United States settled their claims in 1996, with the company paying the United States another $129 million. In addition, there were over two thousand claims made against the company for personal injuries and property damage by residents. Almost all of these claims have now been settled for undisclosed amounts.

The early findings indicating that pregnant women and newborns were at risk from exposure to the chemicals at Love Canal have been confirmed by a recent study of residents living near another infamous toxic waste site, the Lipari Landfill in New Jersey. This study found a significantly lower average birth weight among residents closest to the landfill, compared to the general population. These infants were also twice as likely to be born prematurely. Both low birth weight and prematurity are known to contribute to a host of other medical problems later in life.

With some of the proceeds from the settlement, New York State began a long-term study of the former residents of Love Canal to assess the health effects from living near a toxic waste site. An interim report in 2006 found that, consistent with the initial assessments, there was a positive association between women living along the canal during pregnancy and adverse reproductive outcomes, including low birth weight and congenital malformation. In addition, the study found a higher ratio of female to male births among these women, noting a similar finding to those exposed to dioxins at Seveso, Italy.

Debbie Cerrillo now lives in northern New York and eschews any publicity. Peter Bulka died several years ago. Lois Gibbs founded an important environmental organization, the Center for Health, Environment & Justice, and continues to fight aggressively for the protection of the environment.

The other families that were affected by the Love Canal disaster have tried to move on with their lives. During the relocation, some families were moved to a nearby trailer park. Several years ago it was discovered that the trailer park was on top of a toxic waste site. The people in the trailer park were relocated once again. Another family relocated from Love Canal to Connecticut, only to discover in 1993 that they were two blocks away from a toxic waste site with dangerously high levels of asbestos, lead, and PCBs.

Recently the New York State Legislature dissolved the agency that had been established to redevelop Love Canal. Renovated homes with well-kept lawns, bicycles, toys, and other signs of family life fill the area north of Colvin Boulevard, now named Black Creek Village. The canal is now a grass-covered mound bordered by mature trees with a small building off to one side, all of it surrounded by a chain-link fence. At first glance, it appears to be an inviting play area. Yet underneath the grassy mound, thousands of tons of toxic materials are undergoing chemical treatment before being discharged to the local wastewater treatment plant. East of the site lie abandoned homes and wooded, vacant lots where homes once stood, with grass growing on what remains of the sidewalks and streets. The former canal and the abandoned area will remain uninhabitable for the foreseeable future, serving as a memorial to our failure to protect our environment from toxic chemicals, and to the courage and fortitude of Lois Gibbs, Debbie Cerrillo, Peter Bulka, and the others who taught us how not to forget those who are deeply and personally affected by that failure.

THREE MILE ISLAND, PENNSYLVANIA

1979

Three Mile Island, located in the Susquehanna River of Pennsylvania, is surrounded mostly by farming and small residential communities, such as Middletown, Goldsboro, and Newberry. The island is not far from the tranquil Pennsylvania Dutch countryside of Lancaster County. It is also the location of a nuclear power plant.

Two nuclear power units were constructed at Three Mile Island, or TMI, by the Metropolitan Edison (Met Ed) Company: Unit 1 opened in 1974; Unit 2 opened in December 1978. Both units employed pressurized-water reactors. Within the reactors nuclear fission produced heat that converted water to steam, which in turn powered turbines that produced electricity. The reactor core was protected by a steel case and a large containment structure that was designed to prevent the escape of any radioactive contaminants.

Cooling water is critical to the safe operation of reactors. If the reactor core operates without coolant for an extended period of time, the core overheats and melts the container that surrounds the core. In the early years of the development of nuclear energy, some speculated that in a meltdown, the nuclear fuel could burn through the container, through the underlying soil and bedrock, all the way to China. This scenario became known as the "China syndrome." In fact, *The China Syndrome*, a film that dramatized a nuclear meltdown, opened in theaters across the country on March 16, 1979.

On Wednesday, March 28, 1979, Unit 2 at TMI was undergoing routine maintenance. In the early morning hours, workers were cleaning the condensate polishers, filters filled with resin to ensure that only clean water enters the system. The workers had difficulty flushing a piece of resin from a pipe, and in the process of trying to clean it somehow caused the pumps in the polisher system to shut down. In turn, other pumps that fed water to the system (feedwater pumps) also shut down, thereby shutting down the turbines, and, ultimately, the reactor. The reactor had been operating at close to full capacity, and with the feedwater pumps shut down, there was no water to remove the tremendous amount of heat that remained in the system. The temperature of the reactor rose rapidly, building pressure within the unit. At 4:00 AM, alarms lit up a panel in TMI's control room, and a safety valve opened to release a million pounds of steam into the tranquil surrounding countryside.

A 74-year-old farmer who lived a mile away in Goldsboro was awakened by a terrific roar. He jumped out of bed and looked across the river

A mother and her children walk their dog in Londonderry Township, June 20, 1979, across the river from the cooling towers of Three Mile Island Nuclear Power Plant.
Credit: Photography by Robert Del Tredici, courtesy of the artist

toward TMI and saw a huge cloud of steam flying out of the tower. He, and others like him, went back to sleep, thinking it was an ordinary disturbance, similar to others he had experienced living so close to the nuclear plant. The residents of Goldsboro lived close enough to the plant to hear its alarms go off whenever there was a problem.

Although a light in the control room indicated that the relief valve had closed within seconds, the valve remained open for more than two hours, allowing pressure to drop low enough for the water to vaporize. Faulty signals conveyed more misleading information, prompting operators to take additional measures that drained even more precious water from the system. Without sufficient water to cool the reactor, and with a drop in pressure, fuel elements began to melt. Radioactive water escaped into the basement of the containment building, as well as into an auxiliary building, and hydrogen escaped into the containment dome. The unit was headed for meltdown, and no one at TMI was aware of it.

Finally, someone realized that the relief valve was open, and shut it. The pressure leveled off, but heat remained trapped in the core, which continued to deteriorate. Instruments registered core temperatures above 2,000°F. Operators did not believe that the core could get that hot, and they assumed that the instruments were malfunctioning. It was that hot, and it got hotter, reaching as much as 4,000°F. At 5,000°F, the core would start heading for China.

As the morning wore on, and more alarms sounded in the control room, it became clear that a large amount of radiation had been released into the reactor, the auxiliary building, and elsewhere in the system. A site emergency was declared at 7:00 AM. The radiation readings suggested a possible exposure level for the closest population, in Goldsboro, of 10 rem (Roentgen equivalent man) per hour—a measure of radiation dose. Exposure for workers at the plant was limited to only 5 rem per year, and usual levels within the plant site were 100 millirem per year. This meant that the residents could be exposed at levels thousands of times the normal radiation for the area.

At about 7:30 AM, a general emergency was declared, which meant that an incident had taken place with "the potential for serious radiological consequences to the health and safety of the general public"—TMI's highest alert.

News of the risk could not be confined to the plant after the declaration of a general emergency. Civil defense offices in the area were notified of the emergency, and area residents awoke to radio broadcasts that an accident had occurred at TMI that morning. When they went outside, they could taste metal in the air.

Mayor Bob Reid of Middletown, located across the bridge from TMI, first learned of the accident from his civil defense director. When he called Met Ed headquarters that morning, he was reassured that while there had been an accident, no radiation was released, and there was nothing to worry about. Soon after this conversation, Mayor Reid heard on the radio that radiation had been released. That same morning, Lt. Governor William Scranton III gave a press conference, based on information from Met Ed. Scranton assured the public that everything was under control, that there was no danger, and that no increased radiation had been detected. No sooner had he finished than one of his experts contradicted him, saying it had just been learned that radioactive iodine had been detected offsite. Both the state and Met Ed were quickly losing credibility.

Throughout Wednesday, radiation levels in the dome of the reactor building kept going up, from 800 rem per hour at 7:00 AM to 6,000 rem at 9:00 AM, a level of radiation that would have killed anyone exposed to it within minutes. Hydrogen gas that was generated from the reaction of zirconium and water formed a bubble in the top of the reactor, and other radioactive gases escaped into the atmosphere.

Meanwhile, everyone on TMI was trying to figure out what was happening inside the reactor, and to find a way to cool it down. Donning full protective gear, which included a breathing apparatus, radiation meters, and dosimeters, workers were sent to make adjustments to equipment throughout the system in the ongoing struggle to control the buildup of radioactive releases. At times they had to climb several levels—as much as fifty feet—open a safety valve, and get down quickly, being exposed to radiation the entire time. If their air paks ran out, they had to rip off their masks and catch enough breath to run for safety. They decontaminated by showering, wearing masks so as not to inhale the radioactive particles being washed off. Many "burned out," having been exposed in fifteen minutes of work to the equivalent of their limit of exposure for three months. The experience was enough to terrify even the most experienced workers.

Around midday, while operators were still trying to figure out how to control the situation, a popping sound was heard from within the reactor. Recording equipment showed a sudden, large increase in pressure. No one could pay much attention to it on Wednesday because of other pressing matters. Later it would be of great concern.

By Wednesday evening, the developing story about TMI dominated the national and local television news. Local coverage tended to understate the danger, while national networks competed with stories about the worst possible scenarios. A doctor living in Newberry, two miles from the plant, heard conflicting news reports, some stating there was no radiation danger, others discussing how much radiation might have escaped. His wife was pregnant, and they had a one-year-old at home. They left the area that night, as did others, wary of the conflicting accounts that followed in the immediate aftermath of the accident.

By Thursday morning, the releases seemed to have been controlled, cooling water was restored to the reactor, and tension in the TMI control room eased. But there was little to support Met Ed's optimistic morning pronouncement that less than 1 percent of the fuel elements had failed. The reality was that no one yet knew just how much damage had been done, and only a sample of the contaminated coolant water would give them a firm handle on how much fuel failure there had been, and what dangers still existed.

During the day, contaminated coolant water continued to build up in the basement of the containment building. The operators diluted the contaminated water with large volumes of water from the Susquehanna River and then discharged 400,000 gallons of the coolant, containing xenon-133 and xenon-135, into the river. Residents and officials were furious when they discovered what was going on. While the discharge might not have created a significant threat, it became a public relations nightmare and led to further misgivings on the part of the public and the Nuclear Regulatory Commission (NRC) in their understanding of what was happening. Lt. Governor Scranton visited TMI, as did a host of congressional figures, both to see for themselves and to assure the public that there was no imminent danger.

As Scranton and other politicians were visiting TMI, some of their constituents were trying to get as far away from the site as possible. Rumors spread through the towns and countryside about an impending evacuation, and even a possible meltdown. *The China Syndrome* had opened in the Harrisburg area on Wednesday night. By Thursday, some were no longer willing to wait for further developments. Those who left wondered whether they would ever again see those left behind, and everyone worried about what their children already had been exposed to. While concern and fear were widespread, many residents chose to leave merely out of caution. There was no panic. Not yet, anyway.

On Thursday, operators were finally able to retrieve a sample of the coolant water in the reactor building. They were astonished at the level of radiation they found, which indicated a much higher level of fuel failure than anyone had imagined. Meanwhile, the hydrogen gas bubble remained at the top of the reactor, and slightly increased levels of radioactivity were detected at various locations in the area around TMI. Friday soon became known as "Black Friday," and for good reason.

On Friday morning, a decision was made to address the radioactive gases that had accumulated in a makeup tank, which was filled with water that could be added to the reactor core. Venting to the air was unacceptable, so it was decided that the gas would be transferred to a waste-gas decay tank. Leaks were detected in the line from the makeup tank, causing concern that too much radioactive material might escape into the atmosphere and force an evacuation.

To measure the level of radioactivity escaping during the transfer process, the plant operator directed a helicopter to hover over the cooling towers. As the gases were transferred to the waste-gas decay tank, the helicopter's instruments registered increasing levels of radioactivity, eventually reaching as high as 1,200 millirem per hour, before slowly falling again. Although the reading was high, the ground level of exposure would be diminished by a factor of ten, since the 1,200 millirem was measured at 600 feet in the air. Indeed, a reading at the site fence was only 14 millirem.

When the reading was reported to various agencies, however, it was not made clear that it had been taken at 600 feet above the ground. When the NRC staff received word that a measurement of 1,200 millirem per hour had been taken somewhere at TMI, they concluded, with some panic, that the waste-gas decay tank was leaking at an exposure rate of 1,200 millirem per hour. The NRC decided it had to recommend an evacuation. The staff briefly debated the extent of the evacuation, and finally settled on an area ten miles around TMI.

The NRC conveyed the recommendation to Pennsylvania officials, and it was passed on to other local civil defense officials. Governor Richard Thornburgh was responsible for ordering an evacuation, but he was not convinced it was necessary. After much consultation, Thornburgh ordered schools closed, advised pregnant women and preschoolers to evacuate from a five-mile radius around TMI, just as a precaution, and recommended that others stay indoors and close their windows. Thornburgh gave his order shortly after noon.

The public, however, had reacted even before Thornburgh could make his announcement. A county defense official had jumped the gun and announced in the morning, on the radio, that the governor was going to order a full evacuation. When the governor's announcement came, people were already on the run, and there was a growing conviction that a meltdown was inevitable.

Parents rushed to schools and grabbed their children. Fifth- and sixth-graders wrote their last will and testament as they waited for their parents. Some families got into their cars and drove as far away as possible, leaving their homes unlocked and their doors wide open. Others grabbed their kids, took them home, closed all the doors and windows, and waited.

Telephone lines were overloaded, and few people could call out. Police drove through neighborhoods broadcasting by loudspeakers the school-closing order and the advice issued by Governor Thornburgh. There was a run on banks as people withdrew cash or emptied their safe deposit boxes. Gas stations were jammed. When the decision was made to leave, there was little time to decide what to take. Many felt it was like fleeing a burning house. What do you take in those few moments when you think that you might not ever be able to return? Some took personal telephone books, so they could reach out to their families and friends; some took insurance policies. People took clothes, personal records, birth certificates, pets, baby pictures, a rose bowl that had belonged to a mother or grandmother, stereo and other equipment to be saved from looters, valuable silver that might serve as a down payment on a new home.

In the areas around TMI, 50 to 80 percent of residents fled. Adults watched their children as never before, with every cough, sneeze, or other symptom raising the specter of radiation poisoning, whatever that looked like. The threat was invisible, and therefore all the more frightening.

No sooner had officials dealt with the evacuations than they encountered a larger problem. By Friday, experts realized that the sudden increase in pressure that had taken place on Wednesday was actually a hydrogen explosion. Though the explosion was large enough to destroy a small building, it was a small blip within a 600,000-cubic-foot containment structure with four-foot-thick concrete walls. The hydrogen remained in the reactor, so a much larger explosion was possible. The likelihood of an explosion, and the potential consequences, dominated conversation throughout Friday afternoon and Saturday.

At a press conference, members of the NRC staff stated that a hydrogen bubble could cause a meltdown, but that this was unlikely. Most of the press ran with the story of a possible meltdown. Rumors flew, including tentative plans for the evacuation of a twenty-mile area. At a Roman Catholic service in the area, the priest gave general absolution to the congregation, a blessing for those in mortal danger. On Sunday, President Jimmy Carter visited TMI to demonstrate the unlikelihood of a meltdown.

Within the NRC there were conflicting assumptions about the state of the nuclear reactor. Some assumed that free oxygen could be produced in the reactor vessel and that, when mixed with the hydrogen, an explosion could occur. Others analyzed the situation and doubted whether any free oxygen could be produced in the reactor vessel at all, in which case there was no danger.

It wasn't until Sunday that the NRC reached a consensus that an explosion was highly unlikely. By Monday and Tuesday, the hydrogen bubble was diminishing, and the situation at TMI stabilized. Over the following weeks, the reactor continued to cool down, schools were reopened, and the advisories were lifted.

It took eleven years and nearly $1 billion to clean up TMI. The buildings and equipment were washed with soap and water; low-level radioactive water was decontaminated and discharged into the river. Radioactive material was packed up and sent away for offsite disposal. Reactor Unit 2 will have to be monitored for decades, and then decommissioned.

Over two thousand residents, as well as businesses, filed lawsuits against Met Ed and the manufacturers of the equipment. The economic losses attributable to the explosion were settled for $25 million. The personal injury claims were unsuccessful because the injured plaintiffs could not prove that the accident had released high enough concentrations of radiation to cause their injuries. Those living within five miles of TMI received on average 9–25 millirem of radiation within ten days after the explosion. In comparison, the average annual dose from background radiation in the United States is about 300 millirem, which works out to be about 8.1 millirem for ten days. A twenty-five-year follow-up study in 2004 of more than 35,000 residents who were living within five miles of TMI at the time of the accident found no statistically significant increase in deaths from cancers. There was a slight increase in overall deaths, primarily from heart disease, but it remains unclear to what extent the increase is a result of the stress of living through the disaster and then in the shadow of the two cooling towers in the years since. Most researchers agree that it is too early to determine conclusively what health effects resulted from the radiation exposure.

When the TMI plant was first proposed and constructed, everyone—Met Ed as well as government officials—assured the public that it would be safe, that there was nothing to worry about, that nuclear power was cheap and the technology of the future, and that it was patriotic to build the plant since it would end American dependence on oil from the Middle East. In the spring of 1979, the people in the area around TMI, as well as the wider public, learned that such assurances were unfounded and that this technology carried with it the potential for our own destruction. It is no mystery why nuclear power fell sharply into disfavor following TMI. On March 28, 1979, TMI came perilously close to a frightening meltdown, which would have had devastating consequences for every living thing in the area. Many felt that it was just a matter of time before such a catastrophe happened. Soon it would, at Chernobyl.

TIMES BEACH, MISSOURI

1982

Route 66, a highway that ran from Chicago to Santa Monica, California, has always been part pavement, part myth. At its birth in the 1920s, the road stretched 2,448 miles across eight states, from the conservative farmlands of the Midwest to the glamorous West Coast. The route was designed to connect the main streets of small and large towns along the way, providing access to markets for farm products and a means for Americans to explore the country with their newly acquired automobiles. It also provided an escape to California when land dried up during the Dust Bowl of the 1930s, a journey depicted in John Steinbeck's *The Grapes of Wrath*, where the road acquired the sobriquet "the mother road, the road of flight."

After the interstate highway system was constructed, beginning in the 1960s, Route 66 became obsolete and largely disappeared, physically as well as symbolically. A superhighway replaced the last stretch of Route 66 in 1984. In September 1999, an attempt was made to reconstruct the myth of the road. Route 66 State Park was opened along the Meramec River, twenty miles southwest of St. Louis. The park lies in the Meramec floodplain and covers 409 acres with hiking, biking, and horse trails, and wetlands that attract a broad range of birds, deer, and other game. There is a visitor center along with a small museum of Route 66 memorabilia.

The park is unremarkable, except for a vast mound covered with grass that stands next to the picnic area. The mound, which seems oddly out of place in this landscape, is the grave of the town of Times Beach, Missouri—torn down, bulldozed, and buried. Under the grassy mound lie the remains of houses, mobile homes, and businesses, including the Easy Living Laundromat, the Western Lounge bar, the city hall, and the Full Gospel Tabernacle Church. It was not some mighty natural force that caused such devastation. Instead, it was a small-time waste hauler named Russell Bliss, in league with a company that was trying to save a few dollars on its waste disposal costs.

In the late 1960s, the Northeastern Pharmaceutical and Chemical Company, Inc., or NEPACCO, set up business in a portion of a manufacturing facility near Verona, Missouri, west of Times Beach. The former operator and still owner of the site was Hoffman-Taff, a company that made the defoliant Agent Orange, used by American forces in Vietnam. NEPACCO produced hexachlorophene, an antibacterial agent used in soaps, toothpaste, and hospital cleaners—the same product made at the ICMESA plant in Seveso, Italy. NEPACCO first made trichlorophenol (TCP), and then further refined it to make hexachlorophene. At the end of the distillation process, liquid residues, known as still bottoms, accumulated and were stored in a black 7,500-gallon tank. Disposal of the still bottoms was expensive, and though NEPACCO at first paid an experienced waste company to dispose of the still-bottom residues by incineration at a facility in Louisiana, it later looked for ways to cut costs. When a sales representative at ICP, a local company that sold solvents, heard that NEPACCO was looking for a solution to its high-cost waste disposal problems, the company contracted with NEPACCO to dispose of the still bottoms. ICP knew little about waste disposal, however, and it in turn subcontracted the disposal to Russell Bliss. Bliss operated a waste oil business, collecting used crankcase oil from gas stations and reselling it to refineries, recyclers, and anyone else who would pay for it. ICP charged NEPACCO $3,000 per load and paid Bliss $125 per load. ICP knew the material was potentially hazardous but did not know what was in it. ICP sent a sample of the still-bottom residues to Bliss. He dipped a paper napkin in it, lit the napkin, and concluded that it seemed like a heavy grease.

Bliss, or his workers, drained the NEPACCO waste into a tanker truck, and drove the tanker to his storage facility near Frontenac, Missouri. There the still bottoms were unloaded into storage tanks, which were also utilized to store used crankcase oil. Between February and October 1971,

Bliss picked up six truckloads of still bottoms from NEPACCO, each load containing 3,000–3,500 gallons.

In addition to operating a waste oil business, Bliss kept a stable of Appaloosa show horses. To keep the dust down, Bliss drained the mixture of crankcase oil and still bottoms from his storage tanks in Frontenac and sprayed the material around his horse farm. It worked so well that Bliss began to sell his dust-suppressant services to others, including Shenandoah Stable, near Moscow Mills, Missouri. The owners, Judy Piatt and Frank Hempel, who also kept Appaloosas, paid Bliss $150 to spray the floor of their indoor arena in May 1971. Bliss told Piatt that the material would kill all the flies around the horses. It did more than that.

The night after the spraying, a horse grew quite ill. Within a few days, five more horses lost their hair, developed sores, and became severely emaciated. Sparrows, cardinals, and woodpeckers began to drop from the rafters of the barns. Before long the horses, too, began to die. Piatt blamed the deaths on the spraying, but Bliss denied responsibility, claiming that he had sprayed only used motor oil. To try to stem the flood of deaths, Piatt and Hempel removed a foot and a half of soil from around the arena, but to no avail. Eventually sixty-two horses died or had to be destroyed.

Both Piatt and Hempel suffered diarrhea, headaches, and aching joints. Piatt's six-year-old and ten-year-old daughters also became sick after playing on the floor of the arena. The younger daughter had to be rushed to the hospital on one occasion, and both suffered from gastrointestinal pains and inflamed and bleeding bladders.

A young veterinarian, Dr. Patrick Phillips, who was a graduate student at the time, visited the Piatt stable but could not determine the cause of the illnesses or the deaths of the horses. Because of the unexplained deaths of the horses, and the illnesses of the children, the Missouri Division of Health alerted the federal Centers for Disease Control (CDC) in Atlanta, Georgia. In August 1971, the CDC inspected Shenandoah Stable and collected human and animal blood samples, as well as samples of the soil. CDC representatives also spoke with Bliss, who assured them that he had sprayed his own stable with the same material and that he had not experienced any problems.

Piatt and Hempel took matters into their own hands. In September 1971, they sued Bliss for the injuries and loss of the horses. Starting in late 1971, they also surreptitiously followed Bliss's trucks as waste materials were sprayed or dumped around Missouri. Hempel sometimes wore a wig, Piatt wore a large cowboy hat, and they borrowed different cars to disguise themselves, but Bliss's drivers often recognized them. Piatt and Hempel kept a record of where Bliss sprayed or disposed of materials, keeping up the surveillance for fifteen months.

While Piatt and Hempel followed Bliss, the CDC attempted to identify what might be in the waste oil that could cause such toxic reactions. By late 1972, they were still unable to identify the chemical culprit. Around this time, Dr. Phillips and Piatt heard about the Timberline Stable, where similar problems had occurred, including the loss of twelve horses. The son of the stable owner also contracted a severe skin disorder, chloracne, after playing in the stable. A colleague of Dr. Phillips took samples at Timberline and suffered a burn and then blistering of his face from the soil sample. The CDC was again notified.

In late 1973 and early 1974, the CDC analyzed more soil samples from Shenandoah Stable, and this time the agency found traces of trichlorophenol (TCP), an ingredient in herbicides that causes blistering. When the trace amounts of TCP were administered to the ears of rabbits, they developed the signs of blistering, as expected with TCP. What was not expected was that several of the rabbits died, and autopsies revealed liver damage. This reaction could not be attributed to such small doses of TCP. Something much more deadly was at work.

The CDC ran more complicated tests and discovered that the soil contained tetrachlorodibenzo-p-dioxin (TCDD), more commonly known as dioxin. In fact, the soil samples contained over 30,000 parts per billion (ppb) of dioxin. At this time, though dioxin was known to be deadly to animals, even in small doses, little was known about its effects on humans, and there was no standard for what constituted safe levels of dioxin.

The CDC immediately notified the Missouri Division of Health. Dr. Phillips found Piatt and Hempel at a restaurant and told them the news. He explained what dioxin was, although he himself had only that day learned about it. None of them knew how dangerous dioxin was, only vaguely connecting it with Agent Orange and the Vietnam War.

The authorities began to look for the source of the dioxin. The high concentration of the chemical indicated that it came from an industrial facility. Bliss stated that he got his oil from various sources in Missouri, none of which were industrial sources of dioxin. Dr. Phillips and CDC physicians discovered several facilities in Missouri, including the Hoffman-Taff facility, that could have made Agent Orange or TCP, but none seemed to have any connection with Bliss. Then the investigators located a former supervisor at the Verona plant, who informed them that Bliss had indeed hauled waste from NEPACCO. When they confronted Bliss about the waste hauling he did for NEPACCO, he claimed that he had just remembered the site and was about to call the CDC.

NEPACCO went out of business in 1972, after its main product, hexachlorophene, was banned for most purposes by the Food and Drug Administration (FDA). The ban followed the deaths of thirty-six infants in France who were exposed to high levels of the chemical in talcum powder. Dioxin is an unwanted byproduct of trichlorophenol, a constituent of hexachlorophene.

When the CDC inspected the Verona plant site, NEPACCO was gone, but the tank used to store still bottoms was there, filled with 4,300 gallons of liquid. The CDC tested the material and found dioxin at 343,000 ppb. One CDC representative suggested that there was enough dioxin in the tank to kill everyone in the United States. State and federal authorities, including the Environmental Protection Agency (EPA), focused their efforts on securing and cleaning the Verona site, working with Syntex, the company that had purchased Hoffman-Taff and was therefore responsible for the site. After securing the tank, the most pressing problem was the disposal of the dioxin-contaminated material. One method was to incinerate it, but Missouri did not have any hazardous waste incinerators, and neighboring states threatened to block any attempts to transport the dioxin across state lines. Disposal of the dioxin was delayed until a suitable facility was found.

Dr. Phillips and the CDC investigators also identified another site where dioxin had been sprayed and where several homes were later built. Tests showed high levels of dioxin in the soil. While the CDC recommended that the site be excavated and the people moved, its report also indicated that the half-life of dioxin was one year. Based on the estimate that half of the dioxin would degrade naturally within a year, which was later found to be erroneous, Missouri officials decided to leave the soil intact and not to move anyone.

In 1979, the investigations took another turn. An anonymous tip reported that NEPACCO had buried drums of chemicals on a farm near the Verona plant. Hundreds of drums were uncovered, and dioxin was found in the soil samples. As at the Verona plant site, the first priority was to secure the drums and prevent further discharges before determining how to dispose of the dioxin.

A lack of financial and human resources and insufficient legal authority hindered authorities in their investigation. The federal Resource

Conservation and Recovery Act (RCRA), which was designed to regulate the generation and disposal of hazardous waste, was passed in 1976, but the EPA was slow to enforce the requirements of the new law. Also, RCRA did not address problems associated with old, abandoned hazardous waste sites.

The gap in the law was closed several years later through the passage of the federal Comprehensive Environmental Response, Compensation, and Liability Act, also known as the Superfund law. The Superfund law established a government fund for the investigation and cleanup of abandoned toxic waste sites, with strict liability provisions that allowed the government to recover the costs of the cleanups from the responsible parties. The law was based on the principle that those who threaten public health and the environment through the production and disposal of toxic wastes should be made to pay for the cleanup—the polluter pays principle.

As tough as the law was when it passed in December 1980, it immediately ran into headstrong opposition from the newly elected Reagan administration. Reagan was unsympathetic to environmental issues and immediately set out to diminish the effectiveness of the federal EPA by

After the discovery of dioxin contamination, the residents of Times Beach were blocked from going to their homes and their day-to-day living changed forever.

Credit: © Jim West, courtesy of the artist

cutting resources, delaying regulatory actions, and reducing enforcement. These efforts to undercut the EPA, and the Superfund program in particular, were carried out by Anne Gorsuch, the head of the EPA, and Rita Lavelle, the head of the hazardous waste division. Both Gorsuch and Lavelle joined the EPA from jobs in industries that had been regulated by the EPA. Gorsuch had a reputation from her days as a Colorado legislator as someone who was deeply opposed to federal energy and environmental policies. Many viewed Gorsuch and Lavelle as foxes sent to guard the chicken coop.

The Reagan administration cut EPA funding by 17 percent, and Gorsuch abolished the enforcement office, dispersing the staff into other programs. Soon after Lavelle assumed control of the hazardous waste program, she met privately with industry representatives whose hazardous waste sites were being investigated by the EPA. The meetings led to claims that Lavelle was entering into sweetheart deals with companies to relieve them of the obligation to pay for the multimillion-dollar cleanup of these sites. When the Reagan administration refused to surrender EPA documents to Congress, it was seen as an attempt to hide such deals. There were also reports that the EPA was attempting to lower the standard for dioxin cleanups. This—and the reductions in staffing and resources mandated by Reagan, including laboratories needed to analyze samples—deepened the distrust of both the EPA and the Reagan administration felt by those trying to deal with the dioxin.

After reviewing all of the available records, including Judy Piatt's record of where Bliss had sprayed, an EPA field investigator named Daniel Harris identified numerous sites all over Missouri that might be subject to dioxin contamination. People demanded that the EPA take action to protect those exposed. Rita Lavelle stated repeatedly that no emergency existed, and that since not enough was known about dioxin, more studies were needed before action could be taken. When asked why some of the sites were not fenced, she infamously retorted that fences merely encouraged children to climb over them. Many saw these arguments as attempts to delay the process, as a denial of the seriousness of the dioxin exposure, and as an unwillingness to spend the Superfund money that Congress had appropriated.

The EPA's handling of events in Missouri became an embarrassment in the fall of 1982 when an environmental organization, the Environmental Defense Fund, published a leaked EPA document that listed fourteen confirmed and forty-one suspected dioxin sites in Missouri, and reported that the EPA was going to clean up sites only if the level of dioxin exceeded

100 ppb, whereas the CDC was arguing for cleanups where the dioxin level was only 1 ppb. The town of Times Beach was included on the list. Piatt's records indicated that Bliss's trucks had sprayed his oil mixture on the dirt roads throughout the town. Bliss continued to spray Times Beach from 1972 through 1976. Since the town had the largest population of all the newly revealed sites, it received the most attention. Sampling began in late 1982, and residents in the town soon grew accustomed to people in white moon suits taking samples of the dirt on their streets.

Sampling was completed on December 3, 1982, which was fortunate, because on the following day Times Beach suffered its worst flood in history when the Meramec River overflowed. Residents of the town were evacuated, and it was several days before they could return. Even then, no cars were allowed, and the town was accessible only on foot or by boat. No one under sixteen was permitted to return at that point, and residents were warned to get tetanus shots, not to smoke because of leaking propane tanks, and to obey a curfew.

Many residents attended the town's annual Christmas party at city hall to celebrate the holiday and their safe return after the flood. At the dinner, the residents learned of the results of the samples taken by the EPA. They were shocked out of their holiday cheer. Dioxin had been found in the soil along roads and in backyards. The CDC advised that the people who had not yet returned because of the flood should stay away because of the dioxin, and that those who had returned should get out. Within days, police established roadblocks to prevent access to the town, and people in moon suits returned to take further samples. Times Beach quickly became Missouri's Love Canal.

Despite the growing crisis in Times Beach, officials at the EPA headquarters remained dismissive. Lavelle insisted that there was no emergency. Others closer to the Reagan White House saw Lavelle herself as a disaster in the making. In January 1983 control over events in Times Beach was taken out of her hands.

Further tests conducted by the EPA indicated that dioxin was widespread throughout the town. Officials were uncertain about the health effects of exposure to low levels of dioxin in soil, and even more uncertain about how to dispose of it. The town was situated in a flood plain, and further flooding could spread the contamination. In the end, it was decided that buying the town would be more efficient than relocating the residents for an unknown period while the agencies figured out how to clean up and dispose of the dioxin.

The decision to buy out the town was announced at a press conference on February 22, 1983, by the EPA administrator Anne Gorsuch. The announcement was made to a room full of reporters, while the residents of Times Beach listened to a loudspeaker outside.

Within a few weeks, both Lavelle and Gorsuch were dismissed from the EPA for a variety of reasons, including their handling of Times Beach. Subsequently, Lavelle was convicted of perjury before Congress, of obstructing a Congressional investigation, and of submitting a false statement. She spent four months in jail and served five years of probation.

Meanwhile, the people of Times Beach were stranded. They had to decide whether to stay and wait for the buyout and assume the risks to themselves and their children, or to get out. The authorities had indicated that staying was not safe, but no one could tell them how dangerous it would be to stay. If they chose to leave their homes, they had to find alternative living accommodations and pay for both those accommodations and their Times Beach homes. Businesses in Times Beach were lost, as were the jobs at those businesses. Parents attended countless meetings trying to figure what to do, where to go, for how long, and how to get some financial assistance. Every cough, sore, and fever experienced by the children of Times Beach was watched intently by their parents, always fearing that this was just the first symptom of some unknown disease. Pregnant women worried deeply about the consequences for their babies. For five families that moved away, it was soon discovered that the mobile home park they had moved to was another site that had been contaminated by Bliss. They were forced to move yet again.

The buyouts did not begin until August 1983 and ultimately cost more than $36 million, with the EPA paying 90 percent and the State of Missouri paying 10 percent of the costs. Based on the experience at Seveso, Italy, the state recommended that all the dioxin throughout Missouri be collected and stored in temporary facilities before being incinerated.

Since Times Beach contained over 50 percent of the dioxin in the state, and no one would be living there, it was the logical choice for a new incinerator. Once built, it burned more than 265,000 tons of dioxin-contaminated material, including over 37,000 tons from Times Beach. Syntex was responsible for most of the cleanup at Times Beach and the other sites in Missouri, including the construction of the incinerator, the construction of levees to protect the incinerator and related facilities from flooding, and the demolition and burying of Times Beach itself. By 1997 the cleanup was complete. With the settlement of personal injuries, the costs were close to $200 million.

Judy Piatt and her daughters eventually recovered on their claims against Bliss, IPC, and others. Bliss was prosecuted on a variety of charges, including illegal dumping and tax fraud, and was sentenced to a year in jail on the tax fraud conviction.

People typically visit the Route 66 State Park to pay homage to the famous national highway and perhaps to learn some of its history. Little do they know that the vast mound next to the picnic area, like some prehistoric burial ground, contains the remnants of the lives of some 2,000 people, including their Christmas decorations, their beds, their swing sets, the roofs over their heads—all buried in this spot.

BHOPAL, INDIA
1984

"Run! Gas! Death!" The shouts were heard early one morning in December 1984 in Bhopal, India, after methyl isocyanate (MIC), a deadly gas, exploded from a Union Carbide Corporation (UCC) chemical plant shortly after midnight and continued to discharge for almost two hours. For most, the warnings came too late.

There was a stiff wind from the northwest that night and, since methyl isocyanate is heavier than air, the wind quickly drove the gas to the shantytowns that had grown up around the plant. This area surrounding the plant was overcrowded with families living on small plots of land. Dark, narrow passageways separated the dwellings. The dwellings were constructed of tarpaper, wooden slats, plastic bags, straw, bamboo, and tin. The gas flowed through the open windows and cracks in the loosely constructed shacks. Once inside, it seeped under doors, into bedrooms, and into lungs. Many children and older people—the most vulnerable—died in their sleep. Others woke up coughing, choking, with their eyes on fire. It was dark, with little electricity available in the shacks. They heard noises outside, and when they stumbled to the door, they saw others rushing away, escaping something.

They looked for their spouses, parents, and children. Some were already dead. They grabbed whomever they could and began running to escape the gas. A mother grabbed her infant daughter and ran, coughing and choking the entire time. When she was far enough away to feel some relief from the gas, she looked down and realized that her daughter was dead.

A father awakened to the gas and screams and found his wife already dead in their bed. He panicked and ran, forgetting his children. Only some of his children survived.

Many ran in the same direction as the wind, which carried the deadly poisonous gas over an area of twenty-five square miles. People fell on the streets, choking, vomiting, defecating, and dying. Some foamed at the mouth, others choked on their own blood. Though some of the victims helped others off the ground, many more were trampled to death or run over by cars and trucks. At the nearby railroad station, the stationmaster hurriedly directed an incoming train to keep going and then signaled other stations to keep the trains away. He died at the station.

Methyl isocyanate is highly volatile, flammable, and toxic. It was manufactured at the Union Carbide plant and then stored in underground tanks for use in the production of pesticides. But slow sales resulted in a surplus of pesticides, and the production of MIC was shut down. During the downtime, Union Carbide initiated certain repairs and maintenance procedures, including the cleaning of filters in the pipes that carried the MIC to the process units at the plant.

MIC needs to be kept cool and free of any impurities, including water. So the cleaning of pipes connected to the MIC tanks had to be done with extreme caution, for any water mixing with the MIC could cause an explosive situation. Various safety devices and valves were in place to keep water from entering the MIC storage tanks. On Sunday evening, December 2, 1984, workers began to clean out material that had accumulated on the sides of the pipes. Whether someone failed in their responsibility to keep the valves shut, or whether the valves failed or leaked, water entered MIC Tank No. 610. Compounding this danger, Tank No. 610 was over 75 percent full, which was in violation of safety measures that limited the capacity of the tanks to 50–60 percent. The unfilled space was designed to serve as a buffer to absorb heat if a reaction were to occur in the tank. In addition, the refrigeration unit, which was used to keep the temperature cool, was turned off at the time and the coolant had been drained during the shutdown.

Water leaked into the MIC tank and initiated a chemical reaction that, over several hours, caused an increase in temperature and pressure. Around 11:30 PM on Sunday, workers' eyes began to tear, indicating a leak of MIC. They searched around the MIC tanks and observed that an overhead pipe connected to the MIC tank was leaking. They reported the problem to the supervisor in the control room. The supervisor believed it was only water leaking and, since leaks were common, decided to look into it further after a tea break. At this time, the temperature remained low and the pressure in Tank No. 610 was 10 pounds per square inch (psi), safely within the normal range of 2 to 25 psi.

After the tea break, at about 12:30 AM on Monday, December 3, an operator noticed that the temperature in Tank No. 610 had risen to 77°F, and the pressure had shot up to 55 psi, both very dangerous conditions. Tearing also became pervasive, and workers began to cough. There was an increasing awareness that something very wrong was happening. The operator ran to inspect the MIC storage tank. When he got there, he heard a loud noise, felt intense heat coming from the tank, and saw that the concrete above the tank was cracking. Others at the plant saw a cloud of gas burst out of the plant's stacks. The temperature in the tank was 392°F and the pressure was 180 psi.

Deadly MIC discharged into the open air. Workers tried desperately to stop or control the release. But the safety measures that normally would have helped to control the MIC were inoperative or ineffective. A gas scrubber, using a caustic soda solution, could have neutralized the MIC gas as it was escaping from the system, but it had been turned off during the shutdown. The operator turned the scrubber back on but it never worked. A flare that could have helped control or reduce the gas discharge was also inoperative. The refrigeration system was unavailable since it had been drained of coolant. A water spray that could have reduced the amount of gas escaping did not reach far enough up the stack to be effective. Nothing could be done to stop some forty tons of MIC and other reaction products from exploding out of Tank No. 610 and over Bhopal.

A worker sounded an alarm, but it was a sound heard often from the plant and it was turned off after a few minutes. Few outside the plant heard it and, of those who did, most thought it signaled a shift change. When it was clear that none of the plant's safety systems were of use, the workers still at the plant grabbed oxygen masks or covered themselves with wet cloths. They first checked the direction of the wind, from a large sock tied to a pole, and ran as fast as they could into the wind to minimize their exposure. As they did, they ran past four buses that were designated for the emergency evacuation of residents from areas adjacent to the plant. The buses remained unused. No workers died that night.

Those awakening to this very real nightmare had no idea what to do except to run from the cloud of gas being carried along by the cold wind.

No effective warnings were given, except an occasional loudspeaker on a police van, which just told them the obvious: run for your lives. They had no oxygen masks, and no one told them how to protect themselves.

Dawn broke on a grisly scene. Bodies of infants, children, old people, mothers, fathers, and young people were strewn every which way on roads, in ditches, in doorways, on floors, in beds. A mother and child were found dead, clutching each other. The limbs of the dead were contorted, reflecting sudden, violent deaths; a residue of foam could be seen on their mouths. Carcasses of goats, cows, sheep, and buffaloes were mingled with the dead bodies. Plant life was blackened or dead. Those alive wandered the streets looking among the bodies for their family, neighbors, and friends. The army was activated to help local people gather corpses, and to keep the vultures and dogs away.

While the dead were being gathered, those still alive required help. By early morning, the hospitals in Bhopal were overwhelmed. On the first day, at a hospital with one thousand beds, more than twenty thousand people were treated. Every bed was filled, then every mat, then every space on every floor. Tents were soon set up outside. People were brought to

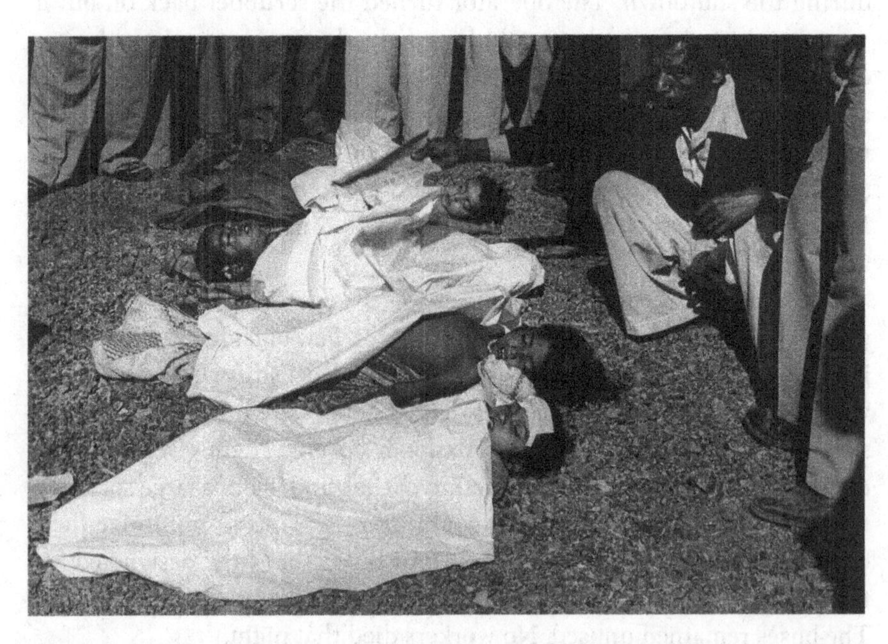

Many of those killed on the night of the gas leak were children. Above, a man pastes identification labels onto the dead children's foreheads before their cremation.

Credit: © Raghu Rai/Magnum Photos

hospitals with burning eyes, ulcers on their corneas, corroded lungs, and inflamed bronchial tubes. Others had difficulty breathing, or were dizzy, unconscious, or in a coma. Throughout the hospitals, children screamed and thrashed, gasping for air to breathe.

Treating the victims was compounded by the lack of information about the nature of the gas to which they were all exposed. When the medical personnel and police asked for information from the Union Carbide plant about the gas, they were told that it was something called methyl cyanate, that it was from an "uncontrolled emission," and that it was not toxic, only an irritant.

Knowing little about MIC, medical staff treated the victims with oxygen and bronchodilators to help their breathing, covered their faces with wet cloths, and washed their eyes. Steroids were given for tissue damage. The lack of disposable needles and the resulting multiple uses of needles increased the potential for infection.

Doctors worried about the spread of disease because of all the dead bodies. Medical staff had more than they could handle, taking care of the thousands of people who showed up at the hospitals. There was scant record of who or how many had died. The corpses were cremated or buried in mass funeral pyres or graves. Animal carcasses were sprayed with lime and salt and buried in large graves. Vultures circled over Bhopal. Over 500,000 people were exposed to MIC gas that night, and some 150,000 suffered injuries, many of which were permanent. Because of the chaotic conditions and the need to bury the bodies quickly in mass graves, the number who died within the first few days still remains uncertain. Officials estimate that more than 3,000 people were killed by the gas, although others estimate that as many as 10,000 were killed. In addition, an estimated 15,000 to 30,000 deaths over the years are attributable to the exposure.

Many aided those in need; others concentrated on placing blame. The lawsuits started in both India and America within a week of the catastrophe and mushroomed after that. During the fight over blame, the causes of the gas release were attributed to factors related to the original establishment and operation of the plant as well as to more immediate events.

India had suffered from chronic food shortages almost since its independence. A major goal of the government was the modernization of agriculture. Increased agricultural production would be achieved through the greater use of pesticides. The goal was reached; between 1956 and 1970 the use of pesticides tripled, and India became a net exporter of grain. A proposed UCC plant in Bhopal stood to contribute to these economic goals.

To accommodate the business, the government waived Indian-ownership requirements when the plant was established. The government insisted, however, that the plant be labor intensive, since Bhopal was located in one of the poorest regions of India, with per-capita income 30 percent below national figures. The low wages in the area suited UCC's bottom line. Ultimately, the establishment of the plant in Bhopal served the interests of the chemical company and of the Indian government, and those interests account, in part, for what happened.

By the early 1980s, the sale of pesticides became highly competitive, and profits fell in many places. At the UCC plant in Bhopal, profits disappeared and losses were incurred for several years. UCC tried to stem the losses, putting the plant up for sale and laying off workers, especially among the supervisory, technical, and maintenance staff. Between 1980 and 1984, staff in the MIC unit was cut by 50 percent, and the plant operated at only 30–40 percent capacity. Losses and cutbacks had taken their toll, and morale at the plant was low. An audit by UCC in 1982 warned of danger from a series of problems at the plant, including the potential danger from overfilling the MIC tanks, a lack of water-spray capability, defective safety valves, and high staff turnover. It was also pointed out that staff had little understanding of MIC and that no emergency plans were in place.

In contrast to the American UCC plant, located in Institute, West Virginia, which relied on computer-run safety systems, the initial pressure to make the Bhopal plant labor-intensive resulted in manually operated safety systems. A leak or danger would be first detected because a worker smelled it. Such problems escaped the notice of the Indian government, largely because it had limited resources devoted to environmental protection and enforcement. No effective citizen watch group existed. The lack of effective oversight allowed the chemical company to minimize expenses wherever possible.

After the explosion, UCC management in America at first claimed that the Indian plant was every bit as safe as the American plant. When it became clear that that was not the case, management blamed the Indians and suggested that it was the Indian operators' lack of training and education and failure to follow company procedures that accounted for the disaster. Implicit in the argument was the belief that a Third World country could not manage the high technology imported by the Americans. UCC management even floated a theory that the release was caused by a saboteur, although no one else gave the story much credence.

UCC attempted to reassure American audiences, especially chemical workers and Congress, that such an event could not occur at UCC plants in America. Concern about a potential accident led to demands for costly safety measures and further regulation of chemical plants. Within two years, based on the events at Bhopal, the federal and some state legislatures passed laws that required companies to inform communities of the nature and volume of chemical substances used in their plants, to report releases of hazardous chemicals from their plants, and to provide advice on medical treatment for exposure to these chemicals.

In response to the management's arguments that the Indians were at fault, others pointed out that UCC had designed the plant, had trained key personnel in America, and had approved operations, including the gas scrubber, flare tower, and water-spray system, all of which failed or were not in use on the night of the disaster. UCC's cost-cutting measures and failure to correct the safety problems, as evidenced by its 1982 audit of the Bhopal plant, contributed to the events.

The forum for resolving the legal disputes was critical. Eventually, an American court held that all claims for injuries and losses should be tried in India, not America. That was a major victory for UCC because it was believed that an American jury would likely base any award on the value of life in America. It was also clear that any litigation in America could be concluded within years, whereas in India it might be decades, and any delay would benefit UCC.

In February 1989, UCC finally agreed to pay $470 million to the Indian government, on behalf of the victims, and to contribute to the construction of another hospital in Bhopal. Each victim would recover between $500 and $2,000 from the settlement. UCC also sold the Bhopal plant. The settlement was widely criticized in India because the government had initially demanded over $3 billion, and victims had demanded $10 billion. The settlement amount was substantially covered by UCC's insurance policies, which meant that the company suffered little as a consequence of the lawsuit. A year later, a new government in India attacked the settlement. India's Supreme Court sustained the settlement, but it allowed criminal actions to be reinstated against Warren Anderson, the former chief executive officer of UCC. Anderson remains a fugitive from Indian justice.

As the legal wrangles continued, so did the effects of the disaster. Approximately 1,000–2,000 animals died from the gas poisoning, and some 7,000 were treated for injuries associated with the gas exposure. Land and crops in the area were contaminated, with 35 of 58 plant species damaged, some even destroyed. Damages to crops have been estimated at $5.2 million, and some land will remain infertile for years. Water impacted by the gas showed increases in carbon dioxide and nitrogen. A week after the disaster, people were hospitalized with gas poisoning as a result of eating fish from contaminated water.

Victims continued to die from the gas exposure and resulting complications for years afterward. Some 60,000 victims were not capable of working full time because of respiratory problems, including bronchitis, pneumonia, asthma, and fibrosis. Others continued to suffer from permanent scarring of the eyes.

By the spring of 1985, an increase in the number of premature births and a decline in birth weight among newborns were reported. Women who had been exposed to the gas experienced an increase in stillbirths and spontaneous abortions, a variety of menstrual problems, and a suppression of lactation. Some women were advised to undergo, and underwent, abortions because of concerns about the potential effects of the gas on fetuses.

Such direct, physical impacts were compounded by a host of psychological disorders—including anxiety, neurosis, and depression—that were triggered by the exposure and the experience. Posttraumatic stress disorder was reported ten years after the disaster. The death of many adult males forced women to seek income-producing work to support their families, a particularly difficult task during an economic decline in a country with a history of deep discrimination against women. In other situations, surviving adult males who lost their wives were forced to give up income-producing jobs to care for the family and household, a situation that was equally disruptive to the family structure.

The most significant long-term environmental effect in Bhopal may have nothing to do with the accident. Greenpeace has reported that the soil and groundwater in the area of the UCC plant are heavily contaminated with mercury, carbon tetrachloride, chromium, lead, and other toxic substances as a result of discharges from day-to-day operations at the plant. Those who survived the gas in the air now have to live with threats from chemicals in their groundwater.

CHERNOBYL, UKRAINE

1986

On Saturday, April 26, 1986, at approximately 1:24 AM, Reactor No. 4 at the V. I. Lenin Atomic Power Station near Chernobyl, Ukraine, exploded.

First, a steam explosion blew a thousand-ton plate off the top of the reactor. Two or three seconds later, a hydrogen gas explosion followed. Massive quantities of radioactive particles and gases were expelled into the atmosphere. Valery Khodomchuk, an operator standing near the reactor core, died instantly. A second operator, Vladimir Shashenka, was close to the reactor hall. He was found alive, but died within hours.

The sound of the explosions brought in the plant's firefighters. Particles of graphite, white hot and radioactive, blew out of the core, setting fire to the roof of the reactor building and igniting dozens of fires in the area. The fire officer in charge immediately sent a coded message to fire brigades in Chernobyl, nearby Pripyat, and the Kiev region. Medical centers in Chernobyl and Moscow were also sent a coded message: "1" (nuclear), "2" (radioactive), "3" (fire), and "4" (explosive danger).

The fire on the roof was of most concern because it could easily spread to Reactor No. 3, which continued to operate. A second reactor explosion was unthinkable, and firefighting efforts were therefore concentrated on the roof. Unfortunately for the firefighters, the roof had been constructed of flammable bitumen. Not only were the firemen directly exposed to radioactive substances, but the boiling bitumen burned through their boots and clothes, eating through to their skin and exposing them further to radioactivity.

Firefighters from nearby areas arrived with little understanding of what they faced. Some kicked, or even picked up, chunks of hot material lying on the ground, not realizing that it was radioactive graphite from the reactor core. While they obviously knew they were fighting a fire at a nuclear power station, the firemen were not told that they were being exposed to high levels of radioactivity. By 6:30 AM on Saturday, the roof and other fires had been extinguished by close to two hundred firemen and eighty engines from the surrounding area, including Kiev. Six firemen died over the following days and weeks.

In the early hours, doctors and nurses from Pripyat rushed to the plant, tending to firemen and plant operators, taking the sick to clinics, and returning to the site to treat others. The doctors and nurses did not know, or did not have time to consider, that the clothing and even the flesh of the exposed firemen and operators emitted radioactivity. Some worked at the scene of the reactor explosion in medical gowns, without any protective clothing. The exposure killed one local doctor and sickened many.

As the fires were fought, policemen and military personnel were brought in from throughout Ukraine. Eventually, more than 16,000 policemen participated in the emergency operations. For their efforts, 57 suffered acute radiation sickness; 1,500 suffered chronic respiratory or digestive problems; and 4,000 exhibited lesser symptoms. A total of 31 people died within weeks from radiation exposure.

For responding to the firestorm and radioactive conditions, and for preventing an even worse disaster, plant operators, firemen, doctors, nurses, policemen, and other volunteers were hailed as heroes. They soon became known as "*likvidatory*," the Russian term for "liquidate." The irony is almost absurd: they both liquidated the fire and were themselves liquidated in the process. For working in such dangerous conditions, they were entitled to extra pay, which came to be known as the "coffin allowance."

To meet certain government target schedules, reactor No. 4 at Chernobyl had been officially cleared for operation on December 31, 1983. While this clearance enabled those involved to claim credit, thereby earning them bonuses and awards, a critical safety test was not performed, although it was mandated before operation could begin. The safety test was finally conducted in April 1986, on the night of the explosion, during a routine

The destroyed reactor after the explosion and fire as seen from the open door of a helicopter.
Credit: ©AP Photo/Igor Kostin

shutdown of the reactor. By then the reactor had been operating for several years, and it was loaded with radioactive fuel.

In power plants, water is heated to produce steam that rotates turbines to generate electricity. In a fossil-fuel electricity-generating plant, the heating source is a furnace and boiler, with coal or oil as fuel. In a nuclear facility, the heating source is a reactor in which energy results from neutron particles colliding with and splitting atoms of uranium-235. The splitting produces new neutrons that collide with the split atoms of uranium, and so on, creating a chain reaction called fission.

An advantage of nuclear reactor power is that one pound of uranium-235 can produce as much heat as 1,500 tons of coal. The major disadvantage is that the process also produces large amounts of highly radiotoxic and volatile radionuclides, including iodine-131, cesium-134, cesium-137, and plutonium. It is critical that there be enough cooling water to prevent the heating process (atomic fission) from overreacting, since too much heat results in a meltdown. The cooling process, as well as other parts of the system, requires electricity to operate, which necessitates a backup power source in case the normal supply of electricity to the system fails.

The backup system for the reactors at Chernobyl relied on diesel generators to supply emergency power for the cooling pumps and other equipment. But these generators needed sixty to seventy-five seconds to reach full capacity to supply power to the pumps. As it happened, that span of time was too long.

The test conducted on April 25 was intended to determine whether a coasting turbine, one that has lost its power source but is still turning, could provide sufficient power to pump coolant through the reactor until the diesel generators kicked in with full auxiliary power. The nuclear reactors used by the Soviet government at Chernobyl had several design defects, including the lack of a containment structure to keep radionuclides within the plant in case of an accident. The safety test was therefore especially dangerous.

The test had been scheduled for earlier in the day, but for unexplained reasons it was delayed until the evening. By then a smaller, less-experienced shift of workers was responsible for the procedure. Operators violated several safety precautions by turning off an emergency cooling system and a safety device that had been designed to shut down the reactor if too much steam pressure were to build up. Other critical errors were made by operators during the test, such as lowering the power level too much and then later withdrawing almost all the rods that controlled the reaction. At several points, the test should have been terminated. Instead, the cooling system was not able to keep the heating process (fission) in check, and the temperature of the reactor core rose to dangerous levels.

At 1:23:40 AM, thirty-six seconds after the test had commenced, an operator pushed the panic button. At 1:24:00, the nuclear explosion occurred.

Several factors hindered attempts by firemen and other liquidators to control the consequences of the explosion: a roof improperly constructed of bitumen; inadequate clothing and equipment (no respirators, for example); fire trucks with ladders that did not reach the roof, forcing firemen to fight the fire from the roof itself; inadequate training; and inadequate monitoring equipment to measure the levels of radioactive contamination to which the liquidators were being exposed. Although plant operators notified officials in Moscow of an emergency at the site in the first hours of the accident, they underestimated, or distorted, the actual conditions and the scope of the danger. Several officials at the plant were later convicted of crimes for violating safety requirements and failing to report accurate conditions.

While people at the plant engaged in a deadly struggle, a new amusement park, with a Ferris wheel, was preparing to open just two miles away in Pripyat. Built in the 1970s for employees of the plant, Pripyat consisted mainly of high-rise apartment blocks. One Pripyat resident who stayed out late fishing on Friday night heard a sound like a clap of thunder and saw the explosion and subsequent column of flames. He continued fishing until late the next day, convinced it was not a reactor explosion, and noting

Nearby Pripyat was evacuated belatedly after the explosion at Chernobyl.

Credit: © Jean Gaumy/Magnum Photos

that the mushroom-like cloud was moving away from him. He was later hospitalized for exposure.

As the day went on, it became hot, so people sought out cool spots, such as the beach, country cottages, or the reservoir adjacent to the plant. Others were tending their garden plots at the edge of a forest on the outskirts of Pripyat. As the day progressed, the green woods turned red from highly radioactive dust. The area later became known as the "Red Forest."

Those who looked toward the nuclear power plant could see that the reactor building was burning, but anyone who asked was told simply that there was a fire. When some tried to warn others that there were serious problems with the nuclear reactor, they were told by neighbors to mind their own business. No one—not the fisherman, not the gardeners in the forest, not the children playing outside, nor any parent—was warned on Saturday about exposure to radioactivity in the air.

At the plant, gases and radioactive material from the explosion formed a plume more than a mile high. Millions of curies of radioactive debris fell within a two-mile radius of the plant; millions more curies took a journey through Europe on a radioactive cloud.

By Saturday night, April 26, the military chemical troops under Colonel General Vladimir Pikalov had assumed command at Chernobyl. Their instruments showed that radioactivity levels were continuing to rise. At one point, General Pikalov ordered his driver to take him closer to the reactor building. No flames were observed, but a fluorescent light glowed above the burned-out building. The general understood what this detail could mean, but he needed to confirm it. He dismissed his driver, commandeered an armored car equipped to measure radioactivity, crashed through the locked gates, stopped at the destroyed building, took measurements, and drove back out. The measurements confirmed his suspicion: the core was probably melting, and the reactor was continuing to spread radioactive material into the air.

Almost a full day had passed before it was established that the graphite reactor core was still burning. This information was conveyed to the special investigation commission set up by the government, and from the commission to the Kremlin. By late Saturday or early Sunday, everyone in the chain of information and decision making knew how catastrophic the accident had become.

Wind shifts throughout that second night caused the radioactive plume to cross over Pripyat, contaminating it through Saturday night and Sunday morning. When officials realized that the core was still burning and releasing radioactivity, they decided to evacuate Pripyat and an area within a radius of six miles of the reactor.

No official evacuation announcement was made until Sunday afternoon, but residents began to hear rumors late Saturday night. Indeed, civil defense staff visited many homes on Saturday night, distributing iodine tablets and instructing people to take them as a precaution. Radioactive iodine attacks the thyroid, a particularly vulnerable and important organ, especially for the young. If iodine-containing substances are taken in sufficient doses, the thyroid becomes saturated and rebuffs the radioactive iodine. Unfortunately, the children had already been exposed to radioactive iodine for a day and a half.

On Sunday afternoon, residents were finally told to pack clothes for three days because they were being evacuated. In less than two and a half hours, over 45,000 people were moved from Pripyat in a twelve-mile-long convoy of buses. It took several more days to evacuate the rural population within the six-mile exclusion zone. Throughout the zone, however, a number of people hid so they would not have to leave their homes or animals. Two elderly women hid for over a month, living on canned food, before being discovered.

On Monday, April 28, at 7:00 AM, the morning shift reported for work at the Forsmark Nuclear Power Plant on the Baltic coast of Sweden, just north of Stockholm. As they did, alarms went off on highly sensitive scintillation counters that had been installed in each nuclear power plant in Sweden. The monitoring equipment indicated a 150-fold increase over normal levels of radioactivity in the air, but the increase did not give rise to panic since nuclear power plants occasionally experience leaks. Nevertheless, the levels detected required that the reactor be closed down and the workers evacuated. An inspection of the plant began, as it was assumed that there was a leak. Yet nothing was found to be wrong with the reactors despite the presence of radioactivity all around the plant.

At 10:00 AM the radiological authority in Stockholm was notified, and an emergency meeting of the board was soon called. Meanwhile, a report came in from a monitoring station fifty miles southwest of Stockholm of a similar dramatic increase in radioactivity. Soon, other power stations in Sweden, Finland, and Denmark reported similar conditions.

By early Monday afternoon, Sweden's National Defense Research Institute had calculated the movements of air masses. An analysis of weather patterns over the previous days indicated that the wind had generally been blowing north from Ukraine, across the Baltic Sea, and then over Scandinavia. The source of the radioactivity was identified as somewhere in Ukraine.

Air samples, which showed the presence of xenon, krypton, high levels of iodine and cesium, and other heavy elements, indicated that the radioactive particles had not come from a nuclear weapon. Most disturbing, the samples contained pure ruthenium, which melts only at a temperature of 4,082°F; there were also more volatile nuclides (including iodine-131, cesium-134, and cesium-137) than nonvolatile nuclides; and there were hot particles in the fallout. These findings indicated the worst scenario: a meltdown at a nuclear power plant. By Monday evening, Sweden had pinpointed the source as Chernobyl.

The detective work was necessary because the Soviet government had not reported a nuclear accident. Even within the Soviet Union, no official announcement had been made. On Monday morning, unaware of the Chernobyl accident, personnel at a nuclear facility in Belarus switched on radiation detection devices and found hot readings, which indicated a leak. They soon discovered that the readings outside were just as high. Another nearby facility also showed high readings. When a call was made to a central station in Minsk, the response was, "This is not your accident."

Only after Sweden informed the international media of Chernobyl did the Russian government acknowledge that a nuclear accident had occurred. On Monday evening, a Russian television program matter-of-factly announced:

> An accident has taken place at the Chernobyl Nuclear Power Plant. One of the atomic reactors has been damaged. Measures are being taken to liquidate the consequences of the accident. Those affected are being given aid. A government commission has been set up.

At Chernobyl, having faced the fire on the roof of the reactor building and dozens of fires around the building, it was now necessary to fight the graphite fire in the reactor core that continued to discharge radioactive material into the environment. Starting on Sunday, April 27, military helicopters flew over the reactor and dropped sacks of sand and clay, boron carbide, lead, and dolomite, in the hopes that these materials would cool the core, generate gas to blanket the fire, and somehow slow or stop the chain reaction. Over the next six days more than five thousand tons of this material was dropped into the reactor hall. It seemed to work because the levels of radioactivity being discharged dropped from 4 million curies on April 27 to 2 million curies on May 1.

The relief these efforts brought was short lived. On May 2, the releases shot back up to 4 million curies. The material dropped into the reactor hall had suppressed part of the fire, but it had not affected the fire from the fission radionuclides, which did not require oxygen for support. A fire still raged in the core. In addition, the material that had been dropped on the fire endangered the structural integrity of the reactor hall floor, beneath which remained the water used to fight the original fires. If the floor gave way, the heated reactor core would crash through and come into contact with the water, and thus set off an explosion that would release the remainder of the uranium, plutonium, and other radioactive materials in the system.

International attention initially focused on the Soviet government but quickly shifted to the weather. A substantial amount of the larger, heavier radioactive particles fell out of the cloud close to Chernobyl, creating frightening risks there. Elsewhere, the concern was with the lighter particles carried in the radioactive cloud. As long as the radioactive emissions remained in the upper air currents and those currents stayed dry, substantially less radioactive contamination would fall out. Eventually the contaminants would be dispersed widely over the planet. But rain picks up radioactive contaminants suspended in the air, with the potentially devastating effect of concentrating the contamination in hot spots.

That is exactly what happened on Tuesday, April 29, when the cloud arrived over central and northern Sweden and Norway. As a result of the fallout from the rain, the Swedish government banned the importation of food from the Soviet Union, advised people in some areas not to drink or use rainwater, and directed farmers to keep milk-producing cows from grazing on grass.

Particularly hard hit by the rain fallout was Lapland. The Lapps, or Sami, had lived on the edge of the Arctic in northern Scandinavia for more than ten thousand years, and some 10 percent of the population still practiced ancient ways of life. The traditional Sami lifestyle was remarkably efficient since it depended heavily on a renewable resource, the reindeer, for food, clothing, and implements.

Although studies had been conducted on the effects of nuclear bomb test fallout on reindeer in the 1960s, there was concern that following the extensive nuclear accident at Chernobyl, the reindeer would be threatened with extinction. Lichen, the reindeer's primary food source, is a unique organism that takes its nutrients from the air. It also does not shed tissue. That April the lichen became a major threat to the reindeer as the radioactive material from Chernobyl quickly built up in the lichen's tissue.

After passing over reindeer herds in Lapland, the radioactive cloud shifted with the wind, first heading west, then south, and then north. Contaminated clouds passed into Austria, where the mountains forced the polluted air higher, cooling it and causing precipitation. At the end of April, radioactivity rained down in the vicinity of Salzburg, Austria, for two days, with the greatest contamination falling on the upper regions of the Alps and on the high pastures, where cattle grazed.

The Austrian government issued warnings about consuming fresh vegetables, and in early May set limits on acceptable levels of iodine-131 in milk. The government also banned the grazing of cattle on affected pastures. When elevated concentrations of fission products were found in rainwater, it was recommended that children not be permitted to play in sandboxes, rain puddles, or meadows.

As quick as the Swedes and the Austrians were to respond, the French moved slowly. By May 1, the cloud had passed over southern France and then north through Alsace, affecting livestock in the region. Calf thyroids in Alsace had significantly elevated levels of radioactivity, as did cheese from goats grazing in high pastures that were subject to heavy rain and fallout. For weeks no action was taken. Later, the French government explained that it hadn't informed the public because it had decided that there was no risk.

After passing over France, the cloud continued its swing north, crossing Belgium and the Netherlands before arriving on Britain's coast. When the British press first disclosed the nuclear explosion, on Tuesday, April 29, the tabloids ran the expected sensational headlines: "Atom cloud horror," "Red nuke disaster," "Scores feared dead. Thousands flee leak."

Despite the press coverage, the British National Radiological Protection Board (NRPB) reassured the public that Britain was safe and that there was no danger from the fallout. But by Thursday, May 1 there was increasing evidence that the radiation was moving toward Britain, and the NRPB went on partial emergency status. British students returning from Kiev were tested for radiation in their thyroids, and the government began to test milk.

Meanwhile, everyone watched the weather forecast. On Friday, May 2, the weather was dry and sunny. The cloud arrived at the southeast coast of England, and by evening it covered the north of England, Wales, Scotland,

and Northern Ireland. For those trying to enjoy a bank holiday weekend, the NRPB again gave assurances that everything appeared to be safe. The board did add that perhaps it would be best not to drink too much rainwater.

While Friday's weather was pleasant in the south, on Saturday heavy rain fell in the north, in Wales and Scotland. Although rain can bring its own depression to Scotland in the best of times, the sheep are keen on its power to produce fresh grass. Luckily, it was lambing season, and the sheep were grazing closer to the farms, where there was less contamination than in the hills and mountains where the rain was heavier. Unluckily, sheep, like reindeer, feed on plants that trap and retain airborne radioactive contaminants.

The cloud departed England on Sunday, May 4. On that same day authorities began to detect increased radiation levels in food supplies. Although the government determined that food with radiation levels above 1,000 becquerels per kilogram (Bq/kg) was unsafe for consumption, they did nothing when sheep grazing in the upland areas of Cumbria, Wales, and Scotland were found to have levels of cesium and iodine as high as 2,450 Bq/kg. Hoping these levels would drop, the government waited six weeks before imposing controls over the movement and slaughter of sheep. Originally, the emergency was expected to last only a short time, a matter of days, perhaps weeks. Instead, it lasted for years.

While the effects of the radiated cloud were witnessed throughout Europe, in Chernobyl authorities were still desperately trying to stop the continuing discharge of radioactivity that followed that first plume of contamination, and to save their own people. With the sudden resurgence of radioactivity on May 2, the exclusion zone was extended to an eighteen-mile radius around the reactor. The majority of people in the first evacuation lived in apartment buildings in Pripyat, but to evacuate the eighteen-mile zone, people had to be relocated from small towns, villages, and rural areas. It took several days instead of several hours to evacuate 90,000 people in this second evacuation. A total of 135,000 people from over 175 villages were permanently uprooted from their homes in the first week. With them went some 50,000 cattle and 9,000 pigs.

The dislocation was traumatic. Many people had lived in the same village for generations, and then, with little advance warning, trucks drove up to the houses and took everyone and their possessions to unfamiliar locations. When a spouse died, the survivor had to get special permission to take the body back home for burial in the family plot.

Older people were resistant, sometimes openly hostile, to relocation, many wailing as if at a funeral. Families often resigned themselves to relocating because they feared for their children. Some doctors advised their pregnant patients to have abortions because of concerns for *in utero* exposure to radiation. Many took their advice. Some groups in the first evacuation from Pripyat had to be relocated a second time because they had been resettled the first time within the eighteen-mile zone. Upon arrival at their new destination, evacuees were measured for radioactivity, which registered on the trousers and hair of many.

Later in May and June, officials temporarily evacuated 64,000 children from Belarus and 25,000 from Kiev to summer camps. An additional 20,000 people were removed from hot spots of contamination detected outside the eighteen-mile zone.

Towns outside of the exclusion zone that showed elevated levels of radioactivity were "deactivated": the top layer of soil from gardens and roads was removed by specially trained chemical army units; the streets were covered with asphalt; and wooden structures were destroyed. Some 500 villages and towns were deactivated through 1986; another one hundred were deactivated in 1987.

On Sunday, May 4, responding to increased levels of radioactivity and at great risk to themselves, workers drained the water from below the reactor hall to prevent a total meltdown. They pumped liquid nitrogen beneath the foundation to freeze the earth in an effort to cool the reactor. Pumping liquid nitrogen into the foundation, and everywhere else it could be done, initially proved effective. A benevolent cold nitrogen cloud embraced the nuclear reactor.

On May 5, radioactive discharges shot back up, to 8 to 12 million curies, close to the level on the first day of the explosions. Many believed a second meltdown was occurring. But later that day, just as suddenly and inexplicably, the levels again dropped dramatically. On May 6, only 150,000 curies were discharged, and levels thereafter continued to decline.

Several theories have been advanced to explain the sudden drop and eventual cessation of radioactive discharge. Based on a recent examination of the buried reactor, it appears possible that the helicopter sorties actually missed the core, which simply continued to burn until it ran out of fuel.

On May 11, Soviet television reported that the danger had passed. On May 14, President Mikhail Gorbachev appeared on television to explain some of what had happened at Chernobyl, and to try to reassure the world that everything was under control. He also condemned Western governments for distorting the events.

Although a collective sigh of relief could be heard, the disaster was not yet over. After the fires were extinguished, radioactive substances continued to be released daily, so attention shifted to permanent containment of the destroyed reactor and the lethal loads of radioactive material. It was also necessary to assess the long-term impact on the environment and the public.

The military was used to remove radioactive dust from the roof of Reactor No. 4. Radioactivity interfered with electronic circuits so that much of the work was done by hand. Soldiers were instructed to start at the edge of the roof in teams of three and to begin counting to ninety. They were told to run to where piles of roofing material had been gathered by others, load the material into wheelbarrows, and dump the material over the side of the building onto the ground below. Once their count reached ninety, they had to drop everything and run back to where they had started; they were then relieved of duty. Those ninety seconds, roughly speaking, constituted the total permissible exposure to radioactivity for five years. It took 3,500 soldiers to remove the debris from the roof.

For several months the government constructed a sarcophagus with some 522,800 cubic yards of concrete to entomb the reactor and its radioactive contaminants, including 2,200 pounds of plutonium. Only after the completion of the sarcophagus, six months after meltdown, did Chernobyl Reactor No. 4 finally stop contaminating the environment.

Further testing revealed that hot spots of contamination covered a much larger area than originally thought. Before the nuclear explosion, the Soviet government had established the permissible level of accumulated doses of radioactivity at twenty-five rem. By 1988, so many villages exceeded the maximum permissible level that the government faced significantly more evacuations and relocations. At the time, the government responded by raising the maximum permissible level to thirty-five rem. After the breakup of the Soviet Union in 1991, however, the governments of Ukraine and Belarus, both of which had been critical of the diminished protection levels, changed the permissible tolerance levels from thirty-five additional rem over a seventy-year lifespan to only seven rem. Between 1990 and 1995 an additional 53,000 people in Ukraine, 107,000 in Belarus, and 50,000 in Russia were evacuated and resettled. Thus, over a ten-year period, more than 325,000 people were required to leave their homes and start their lives over in another place.

In order to determine the long-term health effects from Chernobyl, the former Soviet Union established a registry for over 600,000 people involved in activities related to the cleanup, evacuations, and other activities that might have exposed them to radiation. The hope was to monitor the registered people and assess their long-term health and to learn more about the effects of protracted, low-dose radiation exposures, in contrast to the high-dose, instantaneous exposure suffered at Hiroshima and Nagasaki. Those on the registry undergo yearly medical exams, and they are entitled to social, health, and financial benefits.

The financial costs of Chernobyl and its aftershocks are staggering. By 1993, the costs related to the Chernobyl disaster were $174 billion in Russia, $171 billion in Belarus, and $138 billion in Ukraine. Since the collapse of the USSR, Belarus spends 20 percent of its gross domestic product, and Ukraine spends 12 percent, on problems associated with Chernobyl, particularly in support of the victims.

Radioactive contamination necessitated the removal of topsoil, bushes, and fallen leaves from the exclusion zone, the deactivated towns, and hot spots outside the zone. Not only was cesium found in the soil, but so was plutonium, a particularly radiotoxic substance whose half-life is 24,000 years. In hot spots around the plant, the plutonium was measured at levels sufficient to poison the soil for 1,000 years. Even as far as Kiev, eighty-four miles south of Chernobyl, fallen leaves were removed and buried between layers of clay, and people known as "catastrophists" were observed walking around on bright sunny days covered from head to toe in old clothes, caps, gloves, and stockings.

After removing some 124 acres of topsoil in the area around the plant, officials determined that the amount removed represented less than one tenth of one percent of the contaminated soil. Rather than try to remove all of the contaminated soil, an impossible feat, the decision was made to call the six-mile radius around the plant an "ecological reserve" for the study of the impact of radiation on the environment.

Some arable land was deeply plowed to turn the soil and bury the contaminated layers. While this may have reduced direct exposure, it did not make the soil suitable for agriculture. Perhaps as much as 2.5 million acres of agricultural land in the former USSR will remain contaminated for a century.

Within the exclusion zone, highly radioactive waste was buried in clay-lined pits in some 800 sites. These contaminants, and those in the sarcophagus, threatened the Dnieper and Pripyat rivers, which supply water to 30 million people. To protect the ground and surface water, 140 dams and dikes were built, and a barrier trench five miles long was constructed around the plant at a depth of close to one hundred feet.

Problems also remained in other parts of the world. In the fall of 1986, following the seasonal slaughter of the Sami reindeer, it was discovered that the carcasses contained dangerous levels of radioactive contaminants. A typical Sami family ate reindeer meat six to eight times a week, with a total average weekly intake of two pounds. Given the elevated level of contaminants, each Sami would be subjected to a dose of radiation one hundred times the recommended safe level.

The Swedish government intervened and purchased that year's supply of reindeer meat, but this measure did not solve the long-term problem. Several generations must pass before the lichen is completely cleansed of the radioactive contaminants.

If the authorities enforced a permissible level of 300 Bq/kg of radioactive substances in reindeer meat, a substantial portion of the Sami reindeer would have had to be destroyed. Sami culture depended on the reindeer,

The reindeer in this freezer in Swedish Lapland were unfit for human consumption because the levels of radioactive cesium in their muscle tissue were too high. Workers in the meat-packing plant referred to these carcasses as the "Becquerel Reindeer."

Credit: Photography by Robert Del Tredici, courtesy of the artist

so rather than destroy the culture by destroying the reindeer, Norway simply raised the permissible level for cesium from 300 to 6,000 Bq/kg; Sweden raised its permissible level from 300 to 1,500 Bq/kg. Although at first blush this solution might be seen as absurd or dangerous to the health of the Sami people, there was a precedent to the decision. The permissible level for cesium in Europe at the time was 600 Bq/kg; in the United States it was 1,500 Bq/kg.

The news in England and Scotland was not much better. Emergency controls imposed in June on the grazing and slaughtering of sheep in highly contaminated areas affected over 7,000 farms and 4 million sheep. By 1994, the restrictions still affected 400,000 sheep and 500 farms. By 2002, restrictions remained on the slaughter and distribution of sheep and reindeer in the United Kingdom and some Nordic countries. Some Swedish lakes today still have elevated levels of radioactive elements. Experts have estimated that elevated levels of radioactivity will persist until the next century.

Although the costs of these controls were substantial, the British government did not press claims against the Soviet Union. No legal forum existed to allow it, and apparently some feared that any claim against the Soviets for cross-border contamination would set a precedent, raising the specter of Scandinavian claims against the British for acid-rain damage.

For those directly affected by the contamination the economic impact was severe. The sale of spring lambs constitutes a significant cash crop for English and Scottish sheep farmers, just as the reindeer provide an income for the Sami. Since the radioactive contaminants will continue to affect the reindeer and sheep for a very long time, researchers have been looking for ways to reduce cesium levels to make sheep and reindeer marketable and edible. Some efforts are focused on feeding reindeer a mixture of high potassium and clay minerals in the belief that the potassium will dislodge the cesium and the clay minerals will bind it, allowing the cesium to pass through as waste rather than being absorbed by the gut and then into the meat.

Similar efforts for sheep center on finding an agent that will bind with the cesium, make it indigestible, and allow it to pass in feces. One researcher at Queen's University, Belfast, has experimented with a pigment known as Prussian blue, commonly used in eye shadow, as a binding agent in feed for cows. Another research team has found some success in feeding sheep a byproduct of the soft-drink manufacturing process, citric acid mycelium, to help to flush accumulated cesium-137 from the sheep's body and also block it from being absorbed. The feces, however, is radioactive.

Not surprisingly, the animal kingdom in the former USSR did not fare well. Visitors to the hard-hit Narodichi district of Ukraine, about thirty miles from the Chernobyl plant, reported observing pigs with heads that looked like frogs, a foal with eight legs, and calves born with hare lips or without heads, limbs, or eyes.

In 1986–1987 it was estimated that perhaps as many as 50 million curies of radioactive material were released at Chernobyl. However, recent examinations of the reactor in the sarcophagus have led to a revision. It is now widely accepted that over 150 million curies were released—the equivalent of the fallout from several dozen Hiroshima bombs. Nevertheless, the total release from Chernobyl was less than the release from all the atmospheric nuclear weapons testing that took place from 1945 to 1980.

Chernobyl contaminants were measured over the entire Northern Hemisphere, reaching 5,000 miles away to Japan, as well as to the United States. Close to 400 million people throughout the world were exposed to fallout. Though most were not at significant risk, at the time the scope of the danger was unknown. As the fallout continued to rain down on populations throughout the former USSR and Europe, the catastrophists predicted doom. When the immediate crisis was over and people were not dropping dead in the fields and streets, the scientific community reported confidently that the situation was under control. A 1991 study of the health impact of the disaster sponsored by the International Atomic Energy Agency (IAEA) reported that "no health disorders...could be attributed directly to radiation exposure."

Many others outside the cluster of experts from the energy and nuclear power organizations who conducted the study did not share the confidence embedded in such language. The naysayers soon had professional company. No sooner had the five-year studies offered a rosy gloss than reports began filtering in from the field that sudden, dramatic increases in thyroid cancer had been found in children exposed to Chernobyl fallout. At first these reports were treated with skepticism. Soon the evidence was compelling.

Disputes have arisen regarding the estimated number of cancer deaths attributable to Chernobyl, but there seems to be little question regarding the significant increase in the incidence of thyroid cancer among children in the former USSR. This increase in thyroid cancers was observed particularly in children born before the accident (and thus exposed directly to radiation) and those born within six months after (and thus exposed indirectly *in utero*). Thyroid cancer in children is highly aggressive, and the cancer often spreads to the lymph nodes with metastatic disease on both sides of the neck. Fortunately, if caught early, thyroid cancer can be successfully treated by surgically removing the thyroid gland. After surgery, a high-dose drug replacement therapy is required for the rest of the child's life. Thyroid cancers have occurred frequently enough in the region that the resulting surgical scars around the neck have become known as the "Chernobyl necklace."

In August 2005 the World Health Organization (WHO), the International Atomic Energy Agency (IAEA), the governments of Belarus, Ukraine, and the Russian Federation, and other organizations collectively working as the Chernobyl Forum reported on the health, environmental, and socioeconomic impacts of the disaster almost twenty years afterward. The Forum confirmed that some fifty people died directly from exposure to radiation, including nine children who died from thyroid cancer. Four thousand deaths were estimated to occur from radiation exposure, and another four thousand deaths from thyroid cancer estimated among children and adolescents exposed at the time of the disaster. There was some evidence of elevated levels of leukemia among those exposed, and birth defects for those born later, but it was insufficient to attribute it to Chernobyl.

There is some dispute as to how much radioactive material is contained in the sarcophagus. Some claim that the amount and type of radioactive materials discharged has been underestimated by WHO/IAEA and that the greater quantity created greater risks and effects than WHO/IAEA acknowledge; others support the WHO/IEAE claim that 90 percent of the radioactive material remains within the sarcophagus. So either WHO/IAEA underestimated the amount of material released, in which case the health effects for the already-exposed population are worse, or over 90 percent of the radioactive material remains within a leaking, crumbling sarcophagus, in which case future generations across Europe are at risk from possible releases.

While plans are in place to build a new containment structure around the sarcophagus at a cost of $1 billion, that new layer is expected to last only one hundred years. Some of the radioactive material inside the containment structure will last for thousands of years. And there remains the unsolved problem of the nuclear waste generated by the cleanup. Around Chernobyl hundreds of ditches and trenches filled with radioactive waste are known to exist, but many have not been found or examined since the event.

The final chapter of the nuclear disaster at Chernobyl remains unwritten. Not only does one have to assess the deaths, illnesses, and severe psychological strain resulting from the disaster, which are already identified, but there also are untold costs still mounting from the care of those affected. And there are serious risks of further releases if the containment structure is not constructed, or not constructed safely—a decommissioning issue—and there is the intractable problem of managing the highly radioactive waste from the disaster and from the plant. Of course, it is not only Chernobyl that faces these problems with decommissioning and nuclear waste management—we have seen them at Windscale and possibly at other nuclear facilities.

RHINE RIVER, SWITZERLAND

1986

In Mary Shelley's famous horror story, published in 1818, Dr. Frankenstein observed while boating down the Rhine,

> In one spot you view rugged hills, ruined castles overlooking tremendous precipices, with the dark Rhine rushing beneath; and, on the sudden turn of a promontory, flourishing vineyards, with green sloping banks, and a meandering river, and populous towns, occupy the scene.[1]

This romantic view of the Rhine could not be sustained. Later in the nineteenth century others saw those sloping banks and meandering paths as impediments to the commercial development of the river.

The Rhine River cuts through the heart of Europe, running from high in the Swiss Alps through France and Germany, into the Netherlands, and out into the North Sea. The river's basin covers 77,000 square miles. Some 50 million people live within that basin, and over 8 million rely on the river for drinking water. While other European rivers, including the Volga and the Danube, are longer, the Rhine's position gives it unmatched commercial importance.

In the nineteenth and early twentieth centuries, a series of engineering projects eliminated many bends and curves in the river and turned it into a channel. As a result, the Rhine basin became one of the most densely populated and heavily industrialized areas with steel and other heavy metal manufacturing, as well as 10–20 percent of Europe's chemical industry. With the increased concentration of industry along the river came heavy discharges of fertilizer and pesticides from the agriculture community, polluted runoff from cities along the river, and toxic effluent from industries, especially in the period of economic reconstruction following World War II. As a consequence, the Rhine became notorious as "Europe's sewer."

Shortly after midnight on November 1, 1986, a fire erupted at a warehouse within the Sandoz chemical manufacturing facility, near Basel, Switzerland. Built originally to store machinery and made of corrugated iron, the warehouse had no automatic heat sensors or sprinklers. When the fire broke out, the warehouse contained more than one thousand tons of chemicals in metal drums stacked on wooden pallets. The pallets fueled the flames, and the drums exploded. The fire quickly spread. It took more than five hours for ten fire departments and some 160 firefighters to put it out. In fighting the blaze, millions of gallons of water were poured into

The fire that erupted at the Sandoz warehouse near Basel took five hours to control, and during that time millions of gallons of water contaminated with toxic chemicals were discharged into the Rhine River.

Credit: ©AP Photo/Archiv/Keystone

the warehouse. The chemicals from the warehouse mixed with the water, and though several catch basins were in place to capture runoff from the plant and to prevent discharge into the Rhine, the basins were woefully inadequate to deal with the volume of water generated by the firefight. The toxic brew of water and chemicals flooded Sandoz's sewer system and surface drains, also discharging into the Rhine.

The Swiss trade unions blamed the incident on staff reductions, made after an American efficiency expert was brought in to increase Sandoz's bottom line. The Zurich city police conducted an investigation and concluded that the fire was caused by Sandoz's packaging procedures. Kept in the warehouse was a pigment called Prussian blue, stored in paper sacks on wooden pallets. Plastic was wrapped around the pallets and heat was applied with a blowtorch to shrink it before shipping. If the blowtorch was held over the plastic for too long, the plastic could puncture and the Prussian blue, a highly ignitable substance, could slowly smolder, eventually bursting into flame.

A red cloud of acrid smoke hung over Basel. At 3:00 AM a civil defense siren was sounded, and police cars drove through the city warning everyone to stay inside and to keep their windows shut. The warnings were given only in German, however, and many foreign workers, who spoke little German, opened their windows to find out what was happening. The toxic cloud hung over Basel and nearby French Alsace for a day, prompting many to flee to escape the stench and irritating smoke.

The toxic chemicals formed a red plume in the Rhine more than twenty-five miles long, moving toward the North Sea at two miles per hour. The plume moved inexorably through Switzerland, France, West Germany, and the Netherlands, threatening all fauna, fish, and wildlife in its path, as well as public water supplies. Almost everything along a 150-mile stretch of the Rhine died. Some 500,000 fish, including pike, perch, trout, carp, and over 150,000 eels, were killed by the contaminants. The fish that survived did not survive well: their eyes bulged, their gills collapsed, and their bodies were covered with sores. Fishing businesses along the Rhine lost millions. Children had to be kept away from the river; sheep drinking water from the Rhine died; and swans had to be rescued from the river by environmentalists.

Public water supplies that depended on the Rhine shut down. Some 24,000 villagers in West Germany alone had to get their drinking water from fire trucks for several weeks. The disruption reminded many of World War II, as they carried water home in small containers for drinking and washing. One of the villages affected was Unkel, home of the former West German chancellor Willy Brandt, who described the situation as "Bhopal on the Rhine." Tourists stayed away from the annual grape harvest in Unkel, and cruises along the Rhine, famous for its castles and vineyards, were canceled. German breweries could not use river water for making beer. Fishing was banned; sluices and locks were closed. The Rhine was effectively shut down.

The fire occurred in the early morning hours of November 1, but Swiss authorities did not issue a telex warning to monitoring stations downstream until 7:30 PM on November 2. Even then, they did not know what had been stored in the warehouse or what was in the toxic plume. The European Union's Seveso Directive, drafted in response to the dioxin release in Seveso, Italy, required member states to take precautions to prevent industrial accidents and the ensuing environmental damage, in part by requiring companies to inform local authorities when dangerous chemicals were stored onsite. Switzerland, however, was not a member of the European Union, and thus had no such obligation.

It took Sandoz three days after the spill to identify the thirty-four chemicals that had been stored in the warehouse and discharged into the Rhine. Only then did the authorities, and the public, learn that the toxic plume heading down the Rhine contained over thirty tons of a wide variety of agricultural chemicals, including insecticides, herbicides, fungicides, rodenticides, fertilizers, and other substances toxic to fish and the river's ecosystem. Even more disturbing was the revelation that two tons of organic mercury compounds had also been flushed into the river that night, causing fear of the devastating effects on humans of the mercury poisoning of fish in Minamata, Japan. Sandoz tried to downplay the danger by alleging that the mercury was a less toxic form than that found in Minamata, but many were concerned that it could be transformed into the more deadly form once it entered the river's ecosystem.

When authorities sampled the river, they also found high levels of a weed killer, atrazine. This discovery was perplexing because Sandoz had not included atrazine in its warehouse inventory, and the company neither manufactured nor used the chemical. The explanation for the presence of atrazine in the river came a week later when Ciba-Geigy, another chemical company in Basel, admitted that it had spilled 105 gallons of the chemical into the Rhine the day before the Sandoz spill. It gave assurances that the levels of atrazine were not dangerous, but there was deep suspicion about the accuracy of their claims. Downstream countries found that to account for the levels of atrazine showing up in the river, ten times the reported amount had to have been spilled. The international environmental organization Greenpeace also revealed that Ciba-Geigy had been discharging atrazine into the Rhine for over a year before the Sandoz spill. As for the claim that atrazine posed no risk, recent studies have reported that when frogs ingest atrazine, their hormone systems are disrupted and male frogs grow female gonads. Such studies suggest that atrazine may be a contributing cause of the worldwide decline in frog populations.

Sandoz's failure to alert the public to the plume of toxic waste heading down the Rhine was compounded by its initial dismissive attitudes toward the impact of the spill. When the company was criticized for not having sufficient fire alarms in the warehouse, a Sandoz manager replied, "More fire alarms mean more false alarms." When authorities in Basel were criticized for the Sandoz and Ciba-Geigy spills, a spokesman brushed it off by declaring that "the emission of substances used for agri-chemical production into the Rhine happens frequently."[2] This statement was soon verified when, on November 8, Sandoz leaked another thirty to sixty tons of contaminated water into the Rhine as a result of a broken seal in the plant's underground drainage system.

Despite the economic importance of Sandoz to the city of Basel, the spill and its consequences were deeply embarrassing to the Swiss. The Swiss were leading the fight against nuclear power and the despoliation of forests from acid rain, and they prided themselves on their devotion to cleanliness and on their pro-environmental attitudes. The Sandoz spill sullied that reputation. The failure to warn other countries and downstream users of the water was particularly unsettling since the spill occurred less than seven months after the Chernobyl accident. It is not surprising that many called the incident Chenobale or Chernobasel.

As the toxic red plume flowed downstream on its course to the North Sea through France, Germany, and the Netherlands, the public, in Switzerland and throughout Europe, grew angry. Protests began almost immediately and escalated as the impact from the spill and the cavalier attitude of the companies came to light. A funeral was held in Basel for "Fluvius Rhine. Died Nov. 1." Swiss school children demonstrated; protesters carried signs declaring, "Fish are powerless, we are not," and, "Today the fish, tomorrow us" (*Heute Fische, Morgen Wir!*). A cartoon in a German magazine depicted the mythic Lorelei with her hair falling out because of chemical pollution. Some protesters, called "chaotics," even smashed windows of the chemical companies. As resentment grew, Sandoz representatives and

Basel authorities were pelted with dead eels and bottles of Rhine water at public meetings. Sandoz managers were spat upon, and some received death threats.

The investment community also expressed dissatisfaction with Sandoz. In the six days after the fire, Sandoz stock plummeted, in large part because of the anticipated damages that the company would have to pay. Eventually, Sandoz paid the French government $7.85 million and set up a $6.8 million foundation for environmental research. The company also paid $1.98 million for losses suffered by German interests. To put these damages in context, Sandoz registered sales of $5.1 billion for the year ending December 31, 1985.

Cleanup at the plant site was also costly. A tent-like cover was placed over the warehouse, and a wall was constructed around the building to prevent additional material from discharging into the river. The cover also alleviated the smell of pesticides that had hung in the air for days after the accident. It took more than a month to clean up the warehouse site, and some 2,755 tons of contaminated debris were removed and buried in dumps in Switzerland.

Little could be done about the contaminants flowing down the Rhine except to get any living thing out of the way and avoid the water until the chemicals passed into the open sea and dispersed. However, many chemicals settled into the sediment at the bottom of the river, especially near the point of discharge at the Sandoz plant. These chemicals, including mercury, were a continuing source of toxins to bottom-feeding fish until, several weeks after the spill, divers dragged large hoses along the river bottom and pumped out the contaminated silt. They removed more than 2,200 lbs of chemicals.

The Sandoz spill was widely viewed as the worst instance of pollution to date in a major European river, wiping out decades of work invested in restoring the Rhine. Initial estimates suggested that it would take decades for the river ecosystem to recover from the chemical spill. Not only was there a massive fish kill, but it was expected that the microorganisms on which the fish depended for food would take years to recover.

Fortunately, the microbiology of the river survived better than had been expected, as the steady flow of the river flushed out many of the chemicals, aided by the cleaning of sediment near the plant. Within several weeks of the accident, people started to use the river water again, and within months microorganisms began to multiply in the river, providing food for the depleted fish population. It took several years, however, for the eel population to return to normal levels.

The amount of mercury that spilled into the Rhine was substantial, yet it equaled the amount of mercury discharged into the Rhine every week from all sources. This statistic may have been comforting to those who tried to minimize the effects of the spill, but for those working to restore the Rhine, it merely reflected the fact that by the late 1980s, 40 percent of Germany's total industrial waste stream, as well as 50 percent of urban effluent, was still being discharged into the Rhine. Large spills and daily assaults were taking their toll, and the persistent pollution of the Rhine had to be confronted.

At the time of the spill, an international organization was already in place to deal with the contamination of the Rhine. In the postwar period, the solution to pollution—flush it down the river into somebody else's backyard—was particularly dangerous for the Netherlands, since much of what went into the river ended up downstream. More than half of the Netherlands is below sea level, and the Rhine is the source of 80 percent of Dutch drinking water. Water is also critical for the Dutch flower industry. Everything that happened upstream reverberated in the Netherlands. With the Dutch leading the way, the International Commission for the Protection of the Rhine (ICPR) was founded in 1950 to try to address the contamination.

Although opposition from commercial interests at first limited the ICPR to gathering data about the environmental condition of the river, by the mid-1980s environmental protection efforts had garnered wide support. A 1984 poll found that 59 percent of Europeans supported the loss of some economic growth in exchange for a cleaner environment, similar to findings in the United States. No longer dismissed as a sewer, the Rhine was gradually becoming valued as a natural resource.

The spill galvanized political and citizen support. The Swiss held an extraordinary joint session of both houses of Parliament, normally reserved for national emergencies, to evaluate a report on the Sandoz spill. In the January 1987 German national elections, the Green Party increased its popular vote from 5.6 percent to 8.4 percent, largely owing to public outrage over the Sandoz spill.

The widespread political outrage following the Sandoz spill allowed the ICPR and other groups to take measures that had previously seemed impossible. In 1987 the ICPR adopted the Rhine Action Plan (RAP) to protect drinking water supplies, reestablish fish species that had disappeared, clean sediments, and safeguard the North Sea ecology. The ambitious project set out to inventory thirty major pollutants in the river, including mercury, cadmium, and toxic organic compounds; to identify the sources of these pollutants; and to reduce discharges by 50 percent. Sampling over many years by the ICPR and local water authorities provided a large pool of data on the quality of the Rhine and the chemicals of concern. In addition, monitoring by Greenpeace and other environmental organizations, including the Dutch Stichting Reinwater (Clean Water Foundation), had identified specific sources of pollution.

Building on this data, and with widespread political and public support, by 1995 the RAP did succeed in reducing by 50 percent the discharge into the Rhine for all thirty chemicals. Stiff penalties for those who did not cooperate bolstered the groups' efforts.

The organizers of the RAP chose salmon as a symbol—a poster species with a large public following—for its goal of reestablishing species that had disappeared from the Rhine. In 1900, 150,000 salmon were caught each year in the Netherlands and Germany. By 1920, the number had dropped to 30,000, and by 1960 salmon had disappeared from the Rhine altogether. To reintroduce the salmon, the RAP removed or bypassed the many weirs, dams, and hydroelectric plants that had grown up along the Rhine and its tributaries, which had interrupted natural migratory paths for the salmon. By the 1990s, salmon had begun to reappear in a number of tributaries and sections of the Rhine.

While the Sandoz spill raised the level of support for environmental protection efforts, and the RAP has made significant contributions, much remains to be done. Thirty chemicals have been brought under some control, but there are still some two thousand pollutants that have not. Sediments in the Rhine remain contaminated. Farm fertilizer and urban runoff continue to flow into the Rhine virtually unabated. The salmon have returned to the Rhine basin, but only in small numbers. The RAP remains an important, multinational, cross-border effort to restore a natural resource, but that effort must grow if the Rhine is ever to regain its former beauty.

PRINCE WILLIAM SOUND, ALASKA

1989

For centuries, the Chugach Eskimo have lived along Alaska's Prince William Sound, an enclosed sea with mountainous islands, glacial fjords, and numerous protected bays. The landscape is dominated by a temperate rainforest with the Columbia glacier serving as a dramatic backdrop. The climate is fairly mild for Alaska, with winter temperatures ranging from only 17° to 28°F. The climate and landscape differ from the barren, frozen Arctic conditions often associated with Eskimo life. Although the Columbia glacier once reached to the sound's shoreline, and cruise ships brought tourists right up to its 200-foot-high wall, over the past sixteen years the glacier has retreated eight miles inland, the result of a warming trend in Alaska.

The sound also contains one of the largest undeveloped ecosystems in the United States and one of the continent's largest tidal estuaries, created by the mingling of rivers, tides, and ocean currents. This ecosystem supports a variety of animal life: humpback and orca whales, sea otters and sea lions, salmon, herring, cod and clams, bald eagles, puffins, murres and harlequin ducks, and thousands of other marine mammals and seabirds.

That environment sustained the Chugach Eskimo subsistence way of life for a very long time. Marine mammals and animals provided most of what the Eskimos needed for food, clothing, shelter, and bedding. Flesh was food; bones were used for tools, utensils, and kayak frames; guts were used for thread and rope; oil was used for cooking and lamps.

The term "subsistence" is used to describe harvesting and fishing wild resources for food, fuel, and other necessities. For the Chugach, subsistence flows from and requires communal activities: maintaining equipment (nets, boats); preparing for hunting and fishing expeditions; tanning skins and making clothing, housing, and kayaks from the skins; killing and retrieving the animals; and sharing the bounty. The term refers to a way of life that is not substantially dependent on a market economy. It includes common ownership of resources and the learning of traditional ways, with special skills and understanding to deal with the interdependent relationship of the Eskimos, the animals, and the environment. The term entails attributes of cultural identity, even spiritual aspects, as well as economics. Throughout the twentieth century, and long before, subsistence survived along with economic and technological developments.

World War II brought roads, bases, and a military presence to the area. Statehood came in 1959, following the Alaska Statehood Act of 1958. Under the act, the State of Alaska was entitled to part of the land that was vacant, unappropriated, or unreserved—the public domain. The natives, however, disputed what was being identified as the public domain.

This dispute over title to the land took on added significance with the discovery of oil along the North Slope of Alaska in 1968. Given their way of life, the land dispute was paramount for the natives, but the stakes were also high for those who knew that a proposed pipeline to carry crude oil could not be constructed until title to the land was resolved.

A resolution, of sorts, was established through the Alaska Native Claims Settlement Act (ANCSA) in 1972. The act converted the natives' claim for common ownership of most of the land in Alaska to private ownership of 10 percent of the land, or some 40 million acres, plus a payment of $962.5 million over eleven years. The federal government and the State of Alaska retained the rest.

There was a catch for the natives. Ownership no longer resided with the natives, collectively considered, or with the native tribes of a given area, but with native corporations that were created by the act. Each native was to receive shares in the corporation in their area, with some land reserved for public purposes. Stock ownership would, in the hearts and minds of some of those in Congress, teach the natives the American, capitalist way, and provide them with a living.

Profit-making exploitation of natural resources conflicted, in a most fundamental way, with the traditional Eskimo relationship as caretaker or steward of the natural resources. These resources were converted into units of wealth that were owned by individual stockholders. Moreover, stock in the corporation was issued only to those Eskimos alive as of December 18, 1971. Such a concept was foreign to and indeed potentially destructive of traditional kinship ties through which every member of the village became an owner in common at birth. Since ownership over native land was through stock, it could now be sold, which meant that control could pass to non-natives. The right of an individual to transfer ownership interests conflicted deeply with native communal life. While the ANCSA threatened the Chugach traditional way of life, as long as they had access to the bountiful harvest of unspoiled Prince William Sound and the surrounding seas, the natives could carry on their subsistence hunting and fishing.

After the ANCSA resolved the issue of title to the land, and following the passage of the federal Trans-Alaska Pipeline Authorization Act in 1973, oil exploitation proceeded. The Alyeska pipeline was completed in 1977 and carried crude oil 800 miles from Prudhoe Bay to Valdez on Prince William Sound, the northernmost ice-free port in the United States. The pipeline was constructed only after a protracted, contentious battle over the potential environmental harm that might result from an oil spill, either from a break in the pipeline or from tankers carrying the oil from Valdez. During the heated debates over the construction of the pipeline, the oil industry and its supporters within government assured everyone that all necessary precautions would be taken to protect the environment.

One Chugach native village is located at Chenega Bay in the southwest area of Prince William Sound. It consists of some thirty homes, a school, a Russian Orthodox church, and a community center, but no grocery store. The villagers make their living from the sea. When they awoke on Friday, March 24, 1989, there was much anticipation, as it was spring, and the villagers were ready to collect herring eggs. They expected the arrival soon of the king salmon, which would signal the start of the fishing season. There was also much reflection, and sadness, however, for it was the twenty-fifth anniversary of the Alaska earthquake of 1964.

The earthquake struck on Good Friday, March 27, 1964, registering 9.2 on the 10-point Richter scale. There were seventy-six people living on Chenega Island at the time, just northwest of the current Chenega Bay settlement. Within minutes, when the tsunamis followed, a seventy-foot wave wiped out the village and killed twenty-three people, almost a third of the close-knit community. It was not until 1984, twenty years after the quake, that the original Chenega community was reestablished and the villagers were able to resettle on Evans Island.

When the villagers awoke on March 24, 1989, the morning news brought word of a new disaster: a major oil spill had occurred in the Sound, off Bligh Reef. Reports suggested threats to villages nearer to the wreck, such as Tatitlek. The spill initially seemed far enough away from Chenega not to pose a threat to fishing. That would soon change.

A tanker ship, the *Exxon Valdez*, had left the Alyeska Pipeline Terminal, in Valdez, at the eastern end of Prince William Sound, at about 9:00 PM, Thursday night, March 23, carrying more than 53 million gallons of crude oil, heading for Long Beach, California. The ship was a "supertanker," about the size of an aircraft carrier, built with a single hull. The ship was escorted through the Valdez Narrows by tugboat, and after the Narrows, the ship was on its own, guided by its captain, Joseph Hazelwood, and his crew.

Hazelwood was a fearless and talented sailor who was widely respected by his peers. When he was thirty-two, he became the youngest captain appointed by Exxon. Given his talents and achievements, it was not surprising that the captain was strong-willed and independent, and that he resented Exxon's cost-cutting efficiency experts. When he first started with Exxon, at age twenty-two, there were forty sailors on ships smaller than the Exxon Valdez. The Exxon Valdez started out in 1986 with a crew of thirty-four, but now, in 1989, when Hazelwood was forty-three years old, he had only twenty crew members. Complaining too much might have seemed ungrateful since Exxon had kept him on after he had experienced some drinking problems.

The harbor pilot steered the ship through the Narrows, and Captain Hazelwood retired to his cabin. Although this was a violation of company rules, nothing occurred while he was off the bridge this time. When the pilot had steered the Exxon Valdez through the Narrows and was ready to depart onto the tugboat, Captain Hazelwood was called and returned to the bridge.

Hazelwood then reported to the Coast Guard's Valdez Traffic Center, which tracked the ship on radar, that he was going to divert from the established outbound traffic lane and cross over into the inbound lane to avoid several small ice floes. Ice floes were a common encounter, and Hazelwood had the choice of slowing down to push the ice out of the way or switching over to the inbound lane, traffic permitting, to exit the Sound. Switching

The Exxon Valdez aground on Bligh Reef. After the initial impact, at 12:04 AM on Friday, March 24, 1989, the vessel's chief mate reported that eight of eleven cargo tanks had ruptured.

Credit: Courtesy of the Exxon Valdez Oil Spill Trustee Council

lanes saved time and money. The Traffic Center reported no inbound traffic and concurred with the decision.

Hazelwood gave the third mate a course heading to cross over to the inbound lane, and a second course change, a slight right turn on the approach to Busby Island, that would take the ship out through the inbound lane. Hazelwood repeated the instructions three times to the third mate. Before retiring to his cabin, another violation of company rules, Hazelwood turned on a computer program that automatically and gradually increased the speed of the ship so that it would reach top speed by the time they exited the Sound. Since the vessel was already on course to switch over to the inbound lane, all the third mate and helmsman had to do was make the slight course change once they were in the lane.

The third mate was likely fatigued from long hours of work on the previous day, and an inexperienced mate was at the helm. The third mate plotted on a map the position for making the last right turn, but unfortunately, he did not make the turn. In the absence of the captain, the third mate and helmsman allowed the ship to head straight for Bligh Reef without the necessary course correction to take them out through the inbound lane.

A lookout rushed onto the bridge and reported that they were in trouble and were fast coming up on Bligh Reef. The third mate tried to correct the course, but it was too late. As he called Hazelwood on the phone to report the danger, the ship hit the reef. The chief mate was awakened by the sound of the wreck, and he immediately awakened several crew members with the announcement, "Vessel aground. We're fucked."

When Hazelwood got to the bridge, he tried several maneuvers to try to free the ship from the rocks, but without success. His chief mate reported that eight cargo tanks out of eleven had ruptured, with loss of oil. Oil was shooting forty to fifty feet into the air from one of the tanks.

The wreck occurred at 12:04 AM, on Friday, March 24. At 12:26 AM, Captain Hazelwood radioed the Coast Guard Traffic Center: "We're fetched up, ah, hard aground... and, ah, evidently leaking some oil, and we're gonna be here for a while and, ah, if you want, ah, so you're notified."

"Some oil" turned out to be 11 million gallons, about one-fifth of the cargo. During the first few hours, almost 6 million gallons of oil poured out of the ship. Five million gallons followed before the spill could be stopped. The call to the Coast Guard triggered a flurry of contacts among representatives of the Coast Guard, the Alyeska pipeline, Exxon, the State of Alaska, and many others who became embroiled in the ensuing cleanup.

In determining the causes of the wreck, in assigning blame, Captain Hazelwood was an easy, early target. He had had at least three alcoholic drinks on the Thursday they left Valdez; he had a history of hard drinking and had been arrested several times for drunk driving. On the morning of the spill several persons reported that they had smelled alcohol on Hazelwood's breath. These reports received widespread press coverage. A story line that was often repeated was that of the "drunken sailor running into some rocks." A year later, however, Hazelwood was found innocent of criminal charges of driving a watercraft while intoxicated, and a Coast Guard hearing dismissed the charges of drunkenness and misconduct. Yet, the National Transportation Safety Board investigation concluded that Hazelwood was impaired by alcohol when the ship wrecked. It would appear that he had been drinking, and that he probably was impaired, but that there was insufficient proof to demonstrate that he was actually drunk.

Further investigations revealed other culprits. Fatigue on the part of the third mate and helmsman was considered a factor. The crew had not complied with required rest periods, and Exxon management was blamed for the cost-cutting reduction in crew size. The *Exxon Valdez* was not constructed with a double hull, which would have contained, or at least substantially reduced, the loss of oil. The Coast Guard Traffic Center radar system was out of date and inadequate, and the radar that could have tracked the vessel as far as Bligh Island was out of order. The State of Alaska permitted the harbor pilots to accompany the ships only to the end of the Narrows instead of to the end of the Sound. Both were seen as cost-saving measures by the government that had contributed to the wreck.

Perhaps also relevant was the fact that before the *Exxon Valdez* spill, in the twelve years since the trans-Alaska pipeline was completed, over 8,700 shipments of oil had been transported by tankers, with no major disasters, only minor incidents. As a result, complacency had set in, which served to rationalize all sorts of cost-saving measures.

The causes of the wreck were much easier to sort out than were remedies for the spill or assessment of the damage that resulted. Nobody—neither the Alyeska pipeline consortium, nor Exxon, nor the Coast Guard, nor the State of Alaska—was prepared with the necessary ships, equipment, or personnel to respond quickly to a spill of such magnitude. The Alyeska spill plan relied on a barge with equipment to contain the spilled oil that would respond within five hours. The barge, however, was damaged at the time, and the necessary equipment had been removed. It took more than fourteen hours to prepare the barge, load the equipment, and get to the spill site. A tugboat with lightering equipment—to off-load the oil that was still on the Exxon Valdez—did not reach the wreck until about eleven hours after the spill was reported.

As the barge and tug were dispatched, officials hurriedly tried to collect more lightering equipment and booms (floating barriers), skimmers, and chemical dispersants to contain the oil that had already been spilled. Materials were flown in from as far away as England, and the Valdez airport, accustomed to twenty flights a day, saw some four hundred flights in the first twenty-four hours after the report of the spill.

Stabilizing and off-loading the Exxon Valdez was the priority, in order to prevent some 42 million gallons of oil still on board from pouring into the Sound. No booms were used to surround the spilled oil until after the ships and equipment could be positioned to off-load the remaining oil. It was not until Saturday, March 25, that the ship was finally surrounded by booms. Officials discussed both the use of chemical dispersants and burning of the oil. Both were controversial. Chemical dispersants break up oil into droplets that sink into the water rather than remaining as slicks on the surface that eventually wash up on shore. Opponents argue that the dispersants do not solve the problem but merely move it below the surface where it continues to affect organisms and fish, and that the chemicals are as toxic as the oil. Some countries use dispersants; many use them only as a last resort. At Prince William Sound, the Exxon representatives were pushing for use of dispersants while the fishermen and environmentalists were opposed. Burning the oil as it floats on the water is an extension of the natural volatilization or evaporation of oil when exposed to air. The lighter fractions of the oil evaporate as soon as the oil is spilled and exposed to air. Igniting the slicks burns off substantial amounts of the heavier oil fractions. The problem was that the fire also produced substantial amounts of smoke and soot that carried their own threats; the residents of Tatitlek, for example, were sickened after just one test burn. Again, it was Exxon that favored burning. It was up to the Coast Guard to decide which measures they would take.

By noon on Friday, the oil slick had spread to an area three miles by five miles around the *Exxon Valdez*. The Easter weekend weather was mild and calm, so there was little wind and wave action to spread the oil across the Sound and onto the beaches and shore. When, later that Friday afternoon, the Coast Guard permitted a trial application of chemical dispersants, it was unsuccessful because the calmness of the water prevented the mixing energy required for the chemicals to interact with the oil.

On Saturday people and equipment continued to arrive in Valdez. Exxon, with Coast Guard consent, assumed control of the recovery efforts. By 8:00 AM, the oil had stopped leaking from the *Exxon Valdez*. Mechanical recovery equipment was employed to start recovering the oil. Another chemical dispersant test was conducted. The company thought it worked; the Coast Guard was not convinced. A burn test proved to be more successful.

By Saturday night, the remaining oil had been transferred from the *Exxon Valdez* to another tanker, and about 50,000 gallons had been recovered from the water by mechanical skimming. Some fifty-six vessels, 26,000 feet of boom, and six skimmers were at work in the cleanup operation. The next morning wave action increased so two more runs were tried with chemical dispersants, which were deemed successful. The Coast Guard authorized the use of the dispersants for that evening. Burning was also approved. Although the oil had spread over more than fifty square miles by this point, the situation was improving.

If the cleanup crews had looked up, they would have noticed darkening skies. The wave action that made the dispersants more effective signaled more sinister forces. Around the same time, reports started coming in about oiled and dead birds. On Easter Sunday night, all hell broke loose on the Sound. A spring storm blew in, bringing winds of seventy-three miles an hour. The storm carried a substantial volume of oil all over the Sound, soaking beaches and shores, and covering trees up to forty feet above the ground near the shore. What oil remained on the surface of the water was pushed farther out from Bligh Reef.

More than 10 million gallons of crude oil went flying across the Sound and beyond. Over the next several months, the oil spread over 450 miles from Bligh Reef. More than 1,000 miles of shoreline were oiled. Included in the affected area were a national forest, four national wildlife refuges, three national parks, four state parks, four state critical habitat areas, and a state game sanctuary. For the Chenega, natural resources that were critical to their subsistence were endangered.

The churning action of the wind and waves thoroughly mixed particles of water with the oil, turning much of the oil that was spread around into a black-brown "mousse." While this mousse was less toxic than the unaltered oil, it doubled or tripled the spread of oil around the Sound. It was also heavier than the oil, making skimming more difficult; it was impossible to burn, because of the water content; and it retarded any chemical dispersants. The oil coated rocks and hardened into a tar-like substance. Oil pooled in pockets among the rocks and sank into sediments on cobbled or coarse sand beaches. Chemical dispersants and burning were now practically impossible, containment was no longer possible, and mechanical recovery was made more difficult.

After the Easter storm receded, things were very bleak in Chenega Bay. Elders observed gray cod gathering in shallow water at the boat harbor, an event that had happened only once before, on the day of the quake in 1964. This was seen as a bad omen. Having lost their village twenty-five years ago to the earthquake, the natives now saw their subsistence way of life threatened. They wondered how they would provide for their families and whether the village would have to be abandoned again.

A native chief of another village described the impact at a meeting of mayors in June. The chief acknowledged that much had changed over the years, that they now had schools and many modern appliances, but that such modern advances were put aside and forgotten when the fish came. The water was their source of life and through all the waves of invasions

The Easter Sunday storm blew ten million gallons of crude oil across the Sound, oiling everything in its path. Birds and other wildlife were drenched in oil.

Credit: Courtesy of the Exxon Valdez Oil Spill Trustee Council

and technological change they had never lost their connection to the water. "So long as the water is alive, the Chugach natives are alive," he said. Yet now they were facing "oil in the water. Lots of oil.... Never in the millennium of our traditions have we thought it possible for the water to die. But it is true.... We have never lived through this kind of death."[1] The image of dead water resonated through media coverage.

Within several days of the spill, oiled and dead birds and carcasses of marine mammals appeared everywhere. The geographical scope and impact on the ecosystem of the contamination was disturbing, but that effect was overshadowed by the concrete images of suffering animals carried by extensive television coverage: animals eating toxic kelp; jawless fish; dead birds floating in the water. People observed sea otters scratching their eyes out and beautiful birds pecking holes in their chests, trying to clean off the oil. Sea lions were covered in oil, their eyes decaying.

The sea otter captured much of the media's attention. Before the spill, there were an estimated thirteen thousand sea otters living in Prince William Sound. The animals inhaled the hydrocarbon fumes, ingested the petroleum while grooming, or absorbed the chemicals through their skin. Otters are particularly vulnerable to oil pollution because they do not possess a layer of fat to keep them warm but rely instead on their dense fur to trap air bubbles to insulate them. The oiling destroyed that insulation. Further aggravating the situation, the spill occurred in the spring when the sea otters give birth to pups, so the females were pregnant or nursing when the spill hit. The total killed by the spill has been estimated at three thousand, or almost a quarter of all the otters in the Sound.

Science writer Jeff Wheelwright described the deaths of the sea otters in *Degrees of Disaster*, an account of the oil spill:

> Maneuvering on their backs as is their wont, sea otters ran into oil they didn't see. Their heads and necks blackened, they groomed furiously, rising half out of the water in alarm, but they succeeded only in distributing the oil over their bodies and into their gullets. Their clumped fur let icy water in. The lighter fractions of the crude burned their stomachs and pierced their lungs. In some instances acute emphysema caused air bubbles to bulge out under the skin of an otter's throat like a grisly necklace. Those animals were the first to go.[2]

The harbor seals were also affected, though not as dramatically as the otters. In the weeks after the spill, harbor seals swam through oil and inhaled hydrocarbons; for months they crawled through oil on the rocky shores and rested on oiled rocks and algae. In some areas of the Sound, over 80 percent of harbor seals were oiled. Pups born in May and June became oiled after birth, and in some areas almost all of the pups were oiled. The oil drove the seals crazy. Oiled seals became sick, uncharacteristically tame, and lethargic, with excessive tearing, squinting, and disorientation. At least three hundred of them died. Autopsies documented the same sort of brain damage seen in people who die from sniffing glue or from other solvent abuse.

The Orca killer whales of the Sound fared no better. Killer whales numbered about 350 in Prince William Sound, living in a number of pods, which are stable social groups of maternally related whales. The pods contain up to fifty whales. The members stay together—rarely does one member spend extended time away from the pod—and each pod has its own distinct dialect of echolocation for finding fish and communicating with one another. One pod living in the Sound, identified as the AB pod, contained thirty-six whales in 1988. This pod enjoyed a certain celebrity, being well known to researchers, boaters, and fishers because of the whales' particular friendliness. They often closely followed boats in the Sound.

The AB pod was photographed several days after the spill swimming in oil slicks. Seven of the thirty-six whales have never been seen since and are presumed dead. Among the missing whales were two females that left young offspring. Within a year, another six whales from the AB pod disappeared, including one female that left behind a young calf. All three of the orphaned calves died. Such a mortality rate for killer whales in the Sound was unprecedented. Not only did the AB pod lose a third of its members, but the reproductive potential of the pod was curtailed by the loss of three females and several juveniles. These losses seem to have led to the social disintegration of the pod itself, as evidenced by the departure of one matrilineal group from the AB pod to join another.

The bird population of the Sound suffered the heaviest losses. Tens of thousands of birds were oiled. Workers retrieved more than thirty thousand carcasses from ninety different species, including loons, grebes, puffins, and guillemots. It is believed that as many as three hundred thousand birds may have died from the spill. Wheelwright provides a possible account of how many of the birds may have died:

> The average marine bird weighs just a pound or two. A half-cup of oil, which a 150-pound human might ingest and shrug off, wreaks catastrophic effects. The bird's intestines, bone marrow, liver, kidney and immune system are attacked all at once. The aromatic compounds also interfere with the function of its adrenal and thyroid glands, with its system of converting saltwater to fresh (resulting in dehydration), with its capacity to break down food (resulting in malnutrition).[3]

Fortunately, there were no large fish kills, no fish carcasses to count, apparently because fish readily sense petroleum in water and avoid it. But the concern was that the fish would spawn in an oil-contaminated environment and the fish harvest would be negatively affected.

While the carcasses of dead otters, seals, and seabirds drew public sympathy and outrage, a less visible yet more insidious consequence of the spill was the impact of oil contamination on the priceless ecosystem of the Sound. When oil enters a marine ecosystem, it is taken in by zooplankton, tiny animals that drift in the upper surface water. The petroleum hydrocarbons are stored in the zooplankton's fat reserves or pass through as undigested matter and are discharged as fecal pellets. The hydrocarbons that remain stored in the fat reserves are transferred to fish that feed on zooplankton, and then on to the sea lions, seals, birds, and other animals that feed on the fish. The fecal pellets contaminated with the petroleum hydrocarbons sink to the bottom of the Sound, where they are transformed by bacteria into detritus, a food source for bottom-dwelling sea creatures. Clams and small animals eat the contaminated detritus; crabs, fish, sea otters, and other animals then eat the contaminated clams and smaller animals.

Wildlife rescue operations began as soon as the oiled and dead birds and mammals appeared. A rescue center was established at Valdez. As the oil spread and the problems worsened, other centers were set up farther west on the Sound, as was an animal morgue. Over 1,600 oiled seabirds were taken to rescue centers; half of them survived and were released. More than 350 otters were rescued, and 225 survived. Some argued later that so few otters were saved that it would have been better, certainly cheaper, to let nature take its course, but the graphic pictures of oiled birds and carcasses of otters and sea lions that appeared on national television would not have supported such inaction.

During the first year, 1989, more than 11,000 people and 1,400 marine vessels participated in the cleanup of the oil that covered the shorelines

Oiled beaches were washed with high-pressure hoses. More than one thousand miles of coastline and an area of ten thousand square miles were affected by the spill, much of it inaccessible to cleanup crews.

Credit: Courtesy of the Exxon Valdez Oil Spill Trustee Council

and still remained on the water surface. Inaccessible locations of the rocky shoreline and cobble beaches provided significant challenges. The recovery efforts included skimmers and other materials to collect oil, and booms along shorelines and bays to prevent the oil from reaching shore. Heavily oiled beaches were washed, sometimes with cold, sometimes with warm or hot water, with high-pressure or other hoses. The high-pressure hoses sometimes drove the oil deeper below the rocks where it was more difficult to reach. Chemicals were tried instead of water, but too heavy a dose of chemicals was required to clean a little oil, and the chemicals kept washing off the beaches and into the water. The chemicals became as great a concern as the oil. Mousse, tar balls, and tainted seaweed were removed by hand. Some bioremediation fertilizer was used to break down the oil and remove it. But the standard method was mechanical—people raking, shoveling, wiping, and scrubbing oil by hand from beaches, rocks, and sediments. Whatever was collected was either burned or sent to a hazardous waste site in Oregon.

The cleanup was suspended over the winter of 1989–1990 because of weather conditions, and oily material continued to wash ashore. Fortunately, the winter storms that year removed a substantial amount of oil from the surface of the beaches. In 1990, 1,000 workers returned to 600 shoreline sites, and by 1991, there were only 5 teams cleaning 147 shoreline sites.

As Exxon representatives, government agencies, waves of experts, and almost everyone living on the Sound joined together to stem the contamination, these same parties squared off against each other on a number of legal fronts. Litigation over who was responsible, and for how much money, began almost immediately after the accident. At bottom it was a contest between Exxon and everyone else: the United States, the State of Alaska, commercial fisheries, natives, and anyone who suffered a loss as a consequence of the spill.

After extensive and heated litigation, the governments settled their claims with Exxon in October 1991. Exxon agreed to pay the two governments a total of $900 million in civil damages for restoration of the damaged natural resources, and $250 million in criminal penalties, half of which was forgiven based on several factors, including Exxon's early response to the disaster and its instituting of corrective practices in its shipping fleet. The restoration fund was to be administered by an Exxon Valdez Oil Spill Trustee Council, comprised of three federal and three state trustees. These sums were in addition to the $2.1 billion Exxon had spent on the actual cleanup, $46 million in lost oil and vessel damages, $300 million in settlements with private parties, and $19.6 million in other damages. In total, the spill cost Exxon over $3 billion.

The private litigation went to trial in 1994 and lasted several months. The plaintiffs were commercial fisheries, municipalities, businesses, native subsistence fishermen, and others claiming loss from the spill. The defendants were Exxon and Captain Hazelwood. The jury found that the crash and the damages sustained from that crash were the result of recklessness because Hazelwood had sailed three hours after drinking alcohol; he left the bridge when he should not have; he left the controls in the hands of an unqualified seaman; and he activated the computer program that sped up the ship before making the final turn. Under widely accepted legal principles, Exxon was found responsible for the recklessness of its captain. The jury then found Exxon liable for $287 million in compensatory damages, and a whopping $5 billion in punitive damages. After the verdict for punitive damages, an Exxon lawyer was quoted as saying, "I think it's a case of the jury not appreciating what five billion dollars means." A juror replied, "Well, he can kiss my ass. It is a chunk of change. But eleven million gallons is a chunk of oil."[4] After years of legal appeals, in June 2008 the United States Supreme Court reduced the punitive damage award to $500 million, slightly more than 1 percent of Exxon's profits of $45.2 billion in 2008.

While the natives recovered the value of some of their harvest losses through the litigation, the court, prior to the trial, dismissed their claims for damages to their culture, their subsistence way of life. For such a claim to prevail under principles of public nuisance law, the natives had to have suffered a special injury, or an injury different in kind from that suffered by the general public. The court held that "the right to obtain and share wild food, enjoy uncontaminated nature, and cultivate traditional, cultural, spiritual, and psychological benefits in pristine natural surroundings" was shared by all Alaskans, not just natives, and therefore there was no special injury to the natives.

The Chenegans were not faring much better back at the cleanup. The subsistence harvest areas critical to Chenega Bay suffered the most damage from the oil spill, and for years following the spill the wildlife that survived was suspect in the eyes of the Chenegans because of health concerns. They were told that they could eat foods that did not taste or smell of oil, and that eating foods with low levels of hydrocarbons did not pose a significant additional risk to health, but that they should continue to avoid shellfish from beaches that were still oiled. However, they were also told that no studies existed on the human health effects of consuming oil-contaminated seafood, especially for those whose diet depended heavily on such seafood.

Despite assurances from officialdom, it was clear that animals in their harvest areas were dying from exposure to the oil. The drop-off in the harvest of marine mammals, especially harbor seals and sea lions, staples of the Chenega diet, was substantial. Before the spill, the average mammal harvest provided 145 pounds per person; in 1989, the mammal harvest fell to 3.6 pounds per person; and, in the post-spill period, 1990 to 1994, the harvest rose to only 20 to 35 pounds per person.

Over the first several years following the spill, the Chenegans devoted substantial time to cleanup activities at the expense of subsistence activities. And what subsistence activity there was required more time and greater travel to fish and hunt for uncontaminated wildlife. Finally, cash income from cleanup work distorted the prior balance between subsistence living and cash income from summer cannery work.

Perhaps most damaging, the Chenegans could no longer trust their own judgments about the wildlife or the environment, for the oil created dangers for which their long history and traditions had not prepared them. Since the cultural values of subsistence depend on communal sharing and learning by the young of those ways, such dislocations threatened the subsistence lifestyle. This was especially so since the Chenegans had only recently reestablished that way of life at the relocated Chenega village, and the younger adults were still learning subsistence skills for the first time. As one Chenega villager remarked, "I still hunger for clams, shrimp, crab, octopus, gumboots. Nothing in this world will replace them.... Living in my ancestors' area [I should] be able to teach my kids, but now it's all gone. We still try, but you can't replace them."[5]

By 1992, three years after the spill, most of the oil was gone from slicks and sheens and from the surface of the water in the Sound, as well as from the beaches. In June 1992, the Coast Guard declared the cleanup complete, although it acknowledged that oil remained in spots throughout the Sound, including on some rock surfaces and buried beneath sediments, rocks, and mussel beds. Despite the declaration, the long-term impact on the wildlife populations was unclear.

Initially, the goal of the recovery efforts was characterized as a cleanup. It soon became clear that a total cleanup was impossible, so the recovery was described as a "treatment." Eventually the goal came to be called "environmental stability." While the cleanup efforts were largely curtailed after

June 1992, most of the oil had not been recovered. The U.S. Government Accounting Office has estimated that only about 10 to 15 percent of the oil that is lost in a major spill is ever recovered. But not even that meager proportion was achieved in Prince William Sound. An Alaska agency estimated that only about 3 to 4 percent of the oil spilled from the *Exxon Valdez* was recovered. It is no surprise that even after the official declaration that the cleanup was over, after the government settlement, and after the trial of the private claims, questions remain about the long-term impact of the oil spill on the wildlife populations and the ecosystem of the Sound.

A series of conferences held in 1993 and 1999 published the results of a multitude of studies that had been conducted to determine the impact of the spill. Much of the research reported at these conferences was directed, at least in part, toward the litigation. Studies sponsored by Exxon argued that the projected numbers of dead animals were exaggerated; that certain cleanup techniques (high-pressure hot water washing) did more harm than good; that too much money was spent on animal rescues; and that the shorelines would have recovered if left alone. They also argued that not all of the hydrocarbons in the Sound and wider area came from the *Exxon Valdez* spill, as some came from naturally occurring oil seeps in the wider area, and from increased boat traffic during cleanup activities. Finally, they argued that the death of large numbers of wildlife was by and large irrelevant as long as the populations of the wildlife affected were restored.

Contrary arguments reported that any early exaggerated estimates of dead wildlife were corrected within a short time; that most cleanup techniques (chemical dispersants, burning, high-pressure washing) have downsides as well as benefits; that the public would not tolerate governments' standing passively by as otters and bald eagles and other animals died "naturally" from exposure to oil that had been spilled by a company that makes large amounts of money from the oil; and that shorelines might recover if left alone, but only after many years, during which time the Chenegans and others would be deprived of their subsistence and their environment. Moreover, if there was oil in the Sound environment from other sources, it paled in comparison to the 11 million gallons of spilled *Exxon Valdez* oil.

In 2001, a research team from the National Oceanic and Atmospheric Administration (NOAA) surveyed 4,800 miles of shoreline and ninety-six sites, randomly selected, to determine the presence of any lingering oil along the Sound. They found that oil remained at fifty-eight of the sites. While the surface oil is generally weathered and hardened, like asphalt, the buried or subsurface oil presents greater concern because it remains in place for years, is more liquid and still toxic, and is taken up by animals. That lingering oil is partially responsible for the impeded recovery of certain of the species in the Sound. In 2004, the Exxon Valdez Oil Spill Trustee Council reported that in the fifteen years since the spill there has been little or no clear improvement for the cormorants, harbor seals, harlequin ducks, and pacific herring. Fortunately, there has been substantial progress toward recovery for the AB pod of killer whales and the sea otters, and recovery has been achieved for the bald eagle and pink salmon. Information was inconclusive for other species.

In the 1991 civil settlement with Exxon, the state and federal governments reserved the right to seek up to $100 million in additional damages if the spill caused a substantial loss or decline in species or habitat in the future. In the summer of 2006, the United States and the State of Alaska demanded $92 million from ExxonMobil based on the continued presence of oil along the Sound.

In the meantime, the Chenegans and others have to live with the results of the spill. Even supporters of Exxon have acknowledged that pockets of weathered oil may have adverse local effects for years. The people of Chenega may not be able to prevent another earthquake. But as one bird rescue worker said, "We tend to lead a sheltered life up here because we're out in the wilderness. The power of large corporations has hit home. It's time we started watching what they're doing more closely."[6]

OIL SPILLS AND FIRES OF KUWAIT 1991

The dugong is a strange, wonderful creature, like its closest relative, the manatee. As a mammal, it seems more human than fish. With a rounded head, small eyes, and a large snout, its face is that of a sad walrus. Its tail is like that of a dolphin, and its front fins allow it to move slowly and gracefully through water. The dugong is also known as the sea cow because it grazes on sea grass, devouring fifty-five pounds a day, and weighing up to 880 pounds. In East Africa, the dugong is known as the "wild pig of the coral."

The dugong can live for up to seventy years. One drawback to their long life is that the female raises a calf only every three to five years. It is a fragile, endangered species, and any interruption of the breeding cycle casts a dark shadow over the group's survival.

The waters and sea grass in the Arabian or Persian Gulf, off Kuwait, are a thriving habitat for the dugong. When Saddam Hussein invaded Kuwait in August 1990, he had no concern for the dugong or the environment of Kuwait. Hussein dispatched some 30,000 soldiers and 700 tanks, followed by another 100,000 troops. Kuwait had 20,000 soldiers. The conquest was swift.

Others, however, paid attention to the effects of the war on the environment when Iraq threatened to blow up Kuwait's oil wells and facilities if attacked. At stake was the loss of 100 billion barrels of oil contained in Kuwait's oil fields, representing about 65 percent of the world's reserves. Since oil represented 39 percent of the world's energy consumption, and 95 percent of all transportation energy, the stakes were high.

King Hussein of Jordan triggered the environmental debate in November 1990 when he predicted at an international climate conference that the incineration of even half of Kuwait's oil reserves would dramatically increase carbon dioxide in the air and create a global warming event that would result in food shortages of disastrous proportions. Others who had been exploring the possibility of a so-called nuclear winter predicted the opposite climatic effects. They hypothesized that the smoke, particularly the soot, from the oil well fires would absorb and block out sunlight, causing a cooling of the earth's surface. The darker the soot, the more sunlight would be blocked out.

The possible effect of such climatic changes on the monsoons was of particular concern. If the soot from the fires at the Kuwaiti oil wells rose only into the troposphere, then much of it would wash back to earth with the rain in a short time, causing only severe local threats. If, on the other hand, the soot rose high enough into the stratosphere, it would remain there for years, and would be carried by prevailing easterly winds over Africa and Asia. There it would block the flow of the cooler, moist air from the oceans necessary for the monsoons. Hundreds of millions of people in Asia and Africa depend on the monsoons for their food, and any interruption in the annual rainfall would result in catastrophe.

The debate raged, but the American scientific community was largely silenced. The U.S. administration, under President George H. W. Bush, censored statements by government scientists on issues such as the possible environmental consequences of an American-led attack on Iraqi forces. In January 1991, the U.S. Department of Energy instructed its researchers not to talk to the media about any possible environmental impacts from oil fires or spills. Other researchers were ordered to withhold satellite images of the Gulf region. The Bush administration feared that concern for adverse environmental effects might diminish public support for the anticipated war effort. Public opposition to the Vietnam War was still fresh.

In January 1991, coalition forces, led by the Americans with British and French participation, struck at Iraq and its troops in Kuwait. No sooner had fighting begun than Iraqi forces discharged oil from pipelines and other facilities into the Gulf to deter invasion from the sea. In addition, coalition attacks struck a supertanker and an offshore loading terminal. Altogether, more than 10 million barrels of oil were spilled into the Gulf, contaminating over 398 miles of shoreline.

When the ceasefire was called on February 28, 1991, Iraqi troops carried out Hussein's promise to blow up the oil wells before surrendering or escaping back into Iraq. In all, the Iraqi forces blew up some seven hundred oil wells, many of which caught fire. The oil well fires spewed out up to six million barrels of oil daily, accompanied by sulfur dioxide, nitrogen oxides, soot, and other toxic substances. That amount was roughly equivalent to 10 percent of the world's daily use of oil. The wells that did not catch fire nevertheless poured out oil onto the ground where it formed lakes containing as much as 60 million barrels of oil on the flat, Kuwaiti landscape. Some of the oil from the lakes infiltrated the groundwater; some evaporated or caught fire, adding more pollutants to the air.

When officials were finally able to assess the scope and impact of the oil well fires it became clear that the worst predictions were not realized. Smoke plumes were contained generally within the lower atmosphere and did not loft into the stratosphere to any significant degree. The smoke and soot were not as black as feared, in part because the fires burned more efficiently than had been expected. The wind direction helped. During the worst times, while the oil well fires still burned, the plume of smoke moved in a southeasterly direction, away from areas of concentrated populations, unlike the fog in London or the chemical cloud in Bhopal, India.

The nature of the Kuwaiti oil also helped reduce the extent of the harm. For the most part, it was a light crude oil that contained more volatile aromatic compounds than heavier forms of crude oil, such as the type spilled by the *Exxon Valdez* in Prince William Sound. The lighter crude can be more toxic, but the volatile compounds disperse easily and are less persistent in the environment.

Even if the worst-case scenario of the Kuwaiti fires and spills was not realized, the emissions and consequences were still considerable. Thick smoke clouds from the fires extended over 4,350 square miles, hovering over Iraq, Iran, Qatar, Pakistan, Sri Lanka, India, Bulgaria, and the Soviet Union. In late March, black oily snow fell in the Himalayas in Kashmir. Low levels of soot from the conflagrations were detected 9,000 miles away in Hawaii. Closer to the source of the fires, oil and soot rained on the people, animals, soil, plant life, and water resources of Kuwait. Black-and-white cows turned gray. Sheep lost weight and then lost their wool, finally developing breathing problems before dying. The smoke and soot were thick enough to block sunlight and drop daytime temperatures by as much as 27°F below normal. Smoke and flaring fires could be seen at night and during the day all along the landscape. Tarmac surfaces were too slippery to walk on. Everything was coated with black, oily substances. This lasted for eight months.

The war produced oil well fires and oiled birds.

Credit: © Steve McCurry/Magnum Photos

In February, a string of fires surrounded Ahmadi City in Kuwait, and thick, tarry substances dropped from the sky. While some were able to protect themselves by wearing a mask while outside and using air conditioning indoors during the worst events, many houses had lost their windows from air strikes or were unable to afford air conditioning.

By May, as the oil well fires continued to burn, the soot in the air over Kuwait was equal to the soot produced by 3 million diesel trucks. Breathing the air in Kuwait City was like smoking 250 cigarettes a day. Into the summer, the maternity hospital in Kuwait City had to change the air filters on incubators every two days instead of every six months.

The air over the Kuwait region was dangerous for over six months. The danger from the black air was obvious. Not so apparent was the impact of the oil spills on the Gulf waters. The Gulf is warm and shallow with coral reefs, coastal sea grass, mangroves, and sensitive intertidal zones. This ecosystem produced a catch of 120,000 tons of fish per year for commercial and subsistence fishing. Dolphins, whales, sea turtles, and dugongs were common.

One natural drawback to the warm, shallow Gulf was that it experienced little wave action or tidal energy, which would have loosened oil contamination. It took over three years for the Gulf to flush its waters into the Indian Ocean, in contrast to Prince William Sound in Alaska, which flushed in twenty-eight days. One threat that was recognized immediately was the danger to the desalination plants in neighboring Saudi Arabia, which provided 80 percent of the area's water supply. Booms were installed, and critical financial resources were directed to protecting these desalination plants and other commercial operations along the coast. The same attention, however, was not paid to other environmental impacts of the war.

By the time the war ended, the destruction of the bird populations in the area was considerable. Over a million migratory birds, including several endangered species, visited the region annually. As the birds flew over the area, they mistook flat, reflective surface bodies for water lakes. They landed in what turned out to be oil lakes and were trapped, becoming easy prey for hawks. The hawks, in turn, also became trapped in the oil. Birds were burned when they flew through the oil fire smoke; others were covered with oil from the fires and fell to the ground. A survey in March found that 50 to 75 percent of shorebirds were oiled, and that between 25,000 and 30,000 seabirds were killed by exposure to the oil.

Remarkably, much of the marine ecosystem survived. Most of the oil washed ashore instead of sinking into the Gulf, and that minimized the damage to the coral reefs and sea grass, the haunt of the dugong. Certain precious natural resources, such as the mangroves and green turtles, were protected by immediate, extensive cleaning operations after the war ended. The dolphins in the Gulf escaped extensive harm by simply staying away from the oil.

By July 1991, much of the oil floating on the surface of the Gulf waters was recovered, mainly by mechanical means. Close to 20 percent of the spill was captured, which is fairly high for oil recovery efforts. In Prince William Sound, only about 3 to 4 percent of the oil was recovered, in part because the *Exxon Valdez* oil was heavier and the weather conditions were much more difficult.

Many estimated that it would take a year or two to extinguish the oil well fires, yet it took only nine months, with the last well fire extinguished in November 1991. Firefighters came to Kuwait from all over the world, some 10,000 workers from thirty-four countries. They came in part because the pay was plentiful—Kuwait oil interests poured substantial money into the well cleanup since the burning oil was costing a fortune in lost revenues each day. By June, Kuwait was shipping oil for sale, and by November it was producing 320,000 barrels a day. The loss of oil—to fires, spills, and discharges—cost over $10 billion, and the cleanup cost more than $1.5 billion.

The oil well fires and oil spills in the Gulf waters caused the most visible environmental damage from the Gulf War. Contamination of the land and groundwater was equally destructive, if less noted. Some of the oil was pumped out of the lakes, some seeped into the ground, and some remained as sludge. The oil seeped into the Kuwait desert and spread the contamination over 250 million cubic feet of soil. Oil also seeped into Kuwait's fresh groundwater and contaminated almost half of its freshwater reserves. Over 300 oil lakes remained in Kuwait ten years after the war. Both the soil and groundwater contamination still require remediation, but the resources to address that damage are scarce.

The particulate matter, or PM, emissions were especially problematic for the more vulnerable populations in Kuwait—the very young, the old, and those with respiratory problems. Particulate matter of a certain size, especially particles less than ten micrometers in diameter (PM_{10}), can absorb gases and other pollutants, infiltrate the lungs, and cause severe breathing problems. Before the war the permissible level of PM_{10} in Kuwait was 340 micrograms per cubic meter (ug/m^3) for twenty-four hours. During the fires the PM_{10} levels in Kuwait rose to 610, and even at times as high as

5,400 ug/m^3. These particulates were loaded with the kinds of toxic substances that often accompany oil fires, such as volatile organic compounds, polycyclic aromatic hydrocarbons, benzene, lead, and nickel.

Both the people of Kuwait and the coalition forces were exposed not only to the oil fires and soot, but also to diesel fuel, pesticides, and possibly dust from depleted uranium (DU) ammunition, made with the tailings, or waste, from the uranium-enrichment process. Just as citizens of Kuwait had no protection against the smoke, the soldiers received little training or equipment to protect them against the risks from the smoke and they were reduced to tying scarves or shirts over their mouths and noses.

More than 700,000 U.S. troops were deployed in the war and, upon their return home, tens of thousands claimed to have been sickened by the war. The symptoms reported—fatigue, joint pain, headaches, memory loss, depression, and chronic diarrhea—were common to several medical conditions, including chronic fatigue syndrome, fibromyalgia, and multiple chemical sensitivity. British soldiers who served in the Gulf War experienced similar symptoms.

Returning soldiers firmly believed that they were suffering physically from a "Gulf War disease," while the government attributed the symptoms to posttraumatic stress disorder, a psychological condition. Given the American government's delays in recognizing the medical effects of the use of Agent Orange in Vietnam, many distrusted its denial of a Gulf War disease. In 1998, the U.S. Congress mandated the establishment of the Research Advisory Committee on Gulf War Veterans' Illnesses, which issued its report in November 2008. The committee, after reviewing all the studies completed to date, concluded that Gulf War illness is a serious medical condition affecting 25 percent of the veterans who served in the war and that the illness was not a stress-related, psychological condition. The committee further found strong and consistent evidence that the illness was associated with the use of pyridostigmine bromide, or PB, a pill given to troops to protect them against nerve agents, and with exposure to pesticides used during the war. Research available on the effects of exposure to petroleum smoke and vapors was insufficient, and the committee could not determine whether exposure to the oil fires was also a risk factor for the complex of symptoms associated with Gulf War illness. Finally, there was evidence that high-level exposure to smoke from the fires was a risk factor for asthma among the soldiers, and no association was found between exposure to DU munitions and dust, but again the research was inadequate to eliminate a connection.

Beyond the damage caused by the oil spills and fires, the coalition strikes on Iraq's infrastructure also had significant consequences for the Iraqi people. Power plants were destroyed and the operation of water and sewage treatment plants was interrupted, resulting in water-borne diseases such as cholera. These quickly spread in a population already weakened by malnutrition and lacking medical supplies. In addition to the thousands of direct Iraqi casualties from the war, thousands of Iraqis died from disease, malnutrition, or inadequate medical care as a result of the war. Iraqi children in particular were affected, with a four-fold increase in the incidence of leukemia among children. Some have attributed the cancer to exposure to the coalition's DU munitions, which release radioactive toxic metals in the dust. American tanks that were hit by friendly fire and contaminated with DU dust were brought back to the States and buried as radioactive waste. When the contaminated Iraqi tanks and the DU munitions were left in southern Iraq, they became objects of curiosity for adults and playthings for children.

War is hell because it kills so many in a short time in violent ways. It also causes suffering for those who lose loved ones. We see from the Gulf War that the environmental consequences also wreak havoc on the air, the water, the land, the ecosystems, and the long-term health of those exposed to toxic substances unleashed by war.

DASSEN AND ROBBEN ISLANDS, SOUTH AFRICA

2000

Penguins are charming creatures. Originally called "feathered fish," they are bumbling walkers on land and graceful torpedoes under water. In some ways penguins can seem more human than almost any other animal, walking upright, like us, with their flippers hanging down their sides like arms. Public affection for penguins, however, is a relatively recent development, arising only in the last century with the introduction of penguins into zoos in the northern hemispheres. Once conservationists and the general public got a look at these creatures, penguins earned a worldwide following in the fight for their protection.

Although penguins have long been game for predator seals and other natural enemies, such as native populations, their methods of killing were not systematic enough to diminish penguin populations. Sailors and explorers were the first to menace penguins by killing large numbers for food and oil. Countless expeditions, including those led by Scott, Shackleton, and Cook, participated in the slaughter of penguins and the harvest of their eggs.

The real threat to penguins, however, came with commercial exploitation of their oil and eggs, as well as their guano, which is used as fertilizer. King penguins—trusting animals—were easily herded onto ramps, led to boilers, and pushed in. Each dead penguin yielded about a pint of oil. In the seventeenth and eighteenth centuries, penguins on Dassen Island, off the coast of South Africa, were clubbed to death in such numbers that breeding on the island all but ceased until the twentieth century. When breeding on the island recovered, penguin eggs were gathered for food. From 1919 to 1931, as many as four hundred thousand eggs were taken each year.

In addition to the threat posed by large-scale killing of penguins and the harvest of their eggs, overfishing in some regions depleted penguin food supplies. In other breeding grounds, introduced animals, such as cats, became yet another threat. Even tourists can pose a danger by getting too close to nests, scaring away the parenting penguin and leaving the egg or chick vulnerable to birds and other predators.

By the mid-twentieth century, most penguin harvesting was declared illegal in many countries, and had virtually ceased around the world. But while the twentieth century brought legal protection for penguins, it also introduced a new menace—oil spills. Now it was not humans killing penguins for their oil, but oil killing penguins. Oil spills around the tip of southern Africa were to prove particularly destructive.

We often associate penguins with the Antarctic, where the Emperor penguin holds court, but only four of the seventeen penguin species breed on the Antarctic continent. Most penguins live in less frigid and even temperate climates, such as the African penguins that breed on both Dassen and Robben Islands off the coast of southern Africa. The African penguin, also known as the jackass penguin because it makes a sound like a braying donkey, is about two feet tall, and weighs between 6.8 and 8 pounds. Males are only slightly larger than females and they look very much alike, so it can be difficult to tell them apart.

African penguins spend much of their time feeding in cold waters on sardines, anchovies, squid, and other fish. Breeding occurs once a year. Nests are dug in guano, or sand if the guano has been depleted, and under rocks or bushes. The female lays two eggs, though usually only one survives. The eggs incubate in the nest for thirty-eight to forty-two days, and the parents take turns guarding the nest and feeding the hatchlings regurgitated food.

Penguins have more feathers than most birds, tightly arranged over a layer of down, which they waterproof using oil secreted from a gland beneath their tails. When a penguin dives into water, its feathers compress, trapping air. Feathers, down, and oil provide insulation against the cold waters in which penguins hunt.

Petroleum, however, has a vastly different effect on penguins. After spills, the penguin ingests the oil or breathes in toxins, causing ulceration of the mouth and stomach. The oil can attack the kidneys, liver, and intestines, cause red blood cells to rupture, and may lead to immunosuppressant effects that make the penguin vulnerable to disease. The oil also destroys the insulation of the feathers, which causes the penguin to lose body heat and die from the cold.

By the late 1960s, the largest cargo ships could not pass through the Suez Canal, which substantially increased traffic along the coast of South Africa and around the Cape of Good Hope, also known as the Cape of Storms. And from 1967 to 1975, the Suez Canal was blocked because of the Arab–Israeli conflict, which forced many additional ships to travel around the Cape. Six major oil spills occurred in that period, resulting in injury to or the deaths of hundreds of penguins. Additional spillage from tank washings, bilge-pumping, and fuel from the wreck of cargo ships also posed risks. In 1994, the *Apollo Sea*, carrying iron ore, sank off the South African coast, spilling its fuel and other oil. An estimated ten thousand penguins were oiled, of which some five thousand died.

Such oil spills contributed to the drastic decline of the African penguin population from between 1.5 and 2 million birds at the turn of the twentieth century to about 180,000 birds at the end of the last century, a decline of some 90 percent. It is not surprising that international organizations list the African penguin as a species that requires protection in order to ensure its survival.

Just six years after the *Apollo Sea* disaster, another major oil spill hit breeding grounds on the islands off South Africa, threatening a substantial portion of the remaining population of African penguins in the world.

In June 2000 the Panamanian M.V. *Treasure* was carrying a cargo of iron ore from China to Brazil. Also on board was a supply of 1,300 tons of oil for fuel and other onboard operations. For unknown reasons, a hole developed below the water line during the journey from China to South Africa. The ship sought refuge near the coast to inspect the damage, but when the hole was discovered, the ship had to be towed toward the open sea where repairs could be made. The M.V. *Treasure* did not get far. On June 23, it sank about 6.5 miles off the coast, spilling most of its 1,300 tons of oil into the waters off South Africa near Dassen and Robben Islands, respectively the largest and third largest breeding grounds for African penguins.

Oiled penguins were found coming ashore from the first day of the spill. It was the middle of breeding season, when one parent guarded the nest and fed the chick while the other went to sea for food, so the timing was particularly devastating. Fortunately, the earlier experience with the *Apollo Sea* spill provided some guidance for dealing with the disaster. Several local organizations, including the Cape Nature Conservation and the South African Foundation for the Conservation of Coastal Birds (known as SANCCOB), began to collect oiled birds and to organize a rescue operation. They called on the International Fund for Animal Welfare (IFAW) and the International Bird Rescue Research Center for additional expertise in handling what was expected to be a large-scale effort.

Within a few days it became clear that even these resources were woefully inadequate. While there were some eighteen thousand penguins on Robben Island, the spill was now heading for Dassen Island, the breeding ground for another fifty-five thousand penguins. This meant that more than 40 percent of the African penguin population was at serious risk. If half the penguins died after exposure to the oil, as happened after the *Apollo Sea* incident in 1994, nearly one-quarter of the world's population of African penguins would be lost. More help was urgently needed.

The rescue organizations sent an emergency call over the Internet to zoos and aquariums throughout the world and to anyone with expertise in caring for injured or oiled birds. The response was immediate. More than a hundred experts arrived from fourteen countries, including staff from zoos and aquariums, all at their own expense. In addition to the professionals who flew in from all over the world, more than ten thousand volunteers from Cape Town and beyond stepped forward to help capture, clean, and feed the birds.

Despite the large turnout, the crisis was overwhelming. Besides capturing the oiled birds and sending them to rehabilitation centers for cleaning and feeding, the clean penguins had to be kept from the oiled water, and fed until the oil was cleaned up. Just collecting the penguins would stretch every resource. Feeding this motley and growing crowd seemed impossible.

The oil cleanup began at once. A recently developed Canadian product—Spill-Sorb—was employed to clean the area around the breeding grounds. Made of sphagnum moss, Spill-Sorb is a natural, nonpolluting material that can absorb about ten times its own weight in hydrocarbons.

Portable swimming pools served as temporary homes for the penguins while they were cleaned and fed. Over 140 pools were set up in this disused railway warehouse in Salt River, a suburb of Cape Town, South Africa.

Credit: © Jon Hrusa, courtesy of the International Fund for Animal Welfare

Volunteers hand-scrubbed rocks with wire brushes, then scrubbed them again when waves washed in fresh oil.

The oiled penguins were captured, placed in specially built boxes, and shipped to one of several rehabilitation facilities on the mainland, near Cape Town, and housed in large porta-pools. The birds were sprayed with vegetable oil to loosen the fuel oil, scrubbed with detergents until the oil was gone, and rinsed and dried, a process that took up to forty minutes for each penguin. During the drying process, a veterinarian inserted a tube down the penguins' throats and fed them a multivitamin solution to hydrate and strengthen them. The penguins were then boxed, trucked to another facility, and fed for two to three weeks until their feathers recovered their waterproofing ability and the oil was cleaned from the sea and their breeding grounds.

The penguins were also hand-fed sardines that had been infused with vitamins. Force-feeding was necessary because they had never learned to eat dead food. During the feeding operation, the charm of the penguins was lost on the rescue workers, many of whom were first-time handlers. Volunteers had to stand in the guano-filled pool, grab a bird and hold it between their knees, open its beak, and force a slippery sardine down the penguin's throat, while the penguins beat their flippers, kicked their webbed feet, and bit with their sharp beaks. Between two to three thousand orphaned chicks were also gathered, transported to the mainland, and cared for in several centers. After the oil was removed, the chicks were taken back to their islands by boat and released among other chicks.

Penguins have sharp homing instincts and can swim long distances. It was decided that the only way to save the unoiled penguins was to capture them and ship them five hundred miles up the Eastern Cape. The clean penguins were corralled and placed in special boxes, put on boats or helicopters, and transported to the mainland where they were loaded onto three-tier sheep trucks and driven for eight hours to Port Elizabeth. Once at Port Elizabeth, they were released on the beach. As had been hoped, the penguins made a dash for the sea, and headed home to Robben and Dassen Islands. The journey took several weeks, and along the way the penguins fed and regained the strength they had lost during their confinement and relocation. Eventually, some twenty thousand penguins were released at Port Elizabeth, and by the time they reached home, their breeding grounds and the surrounding sea had been cleaned of oil.

The relocation was no doubt traumatic for the penguins, but this extraordinary event produced three media stars: Peter, Percy, and Pamela. These were the names given to three African penguins fitted with satellite tracking devices. The devices permitted researchers to study the path and behavior of the relocated penguins after they were released at Port Elizabeth and while they swam the five hundred miles to Robben and Dassen Islands. The signals from the devices, which were attached with Velcro, were transmitted to a Web site where the public could follow the course of their journey home.

In the end, the rescue effort saved almost forty thousand African penguins, although some four thousand chicks could not be saved in time and about two thousand captured adults died. The cleanup cost was $7 million, much of which was spent on rescue and rehabilitation. Some of those saved were two-timers, having been oiled, cleaned, and tagged after the *Apollo Sea* spill.

Assessments of the impact of the spill on the African penguin generally indicate that the population has largely recovered. That success is attributed to several factors: the unprecedented rescue and relocation efforts; the expertise of SANCCOB, IFAW, and other local organizations, which had plenty of experience with prior spills; the quick, generous response from experts throughout the world; and the work of more than ten thousand local volunteers. The weather also cooperated; stormy conditions would have altered the outcome substantially. Finally, the insurance company for the M.V. *Treasure* acted promptly and provided interim funding during the rescue effort and made final payment within nine months. Other funding was provided by the World Wide Fund for Nature and by many local companies and environmental organizations.

Despite the success of the rescue, African penguins have suffered a 60 percent decline since 2001. Part of that decline is due to the loss of food sources, particularly anchovies and sardines, which has resulted, at least in part, from global climate change.

BRAZILIAN RAINFOREST

The rainforest is to some a chaotic place. Others have viewed the rainforest as a lush, mysterious wilderness inhabited by Rousseau's noble savages and free of the trappings of civilization, or as the home of El Dorado, the mythical seat of a king drenched in oil and covered in gold dust. Early anthropologists categorized the rainforest as a haven of "primitive" people, the discovery of which could bring academic fame and fortune. In almost all cases, the impulse has been to enter and impose order.

A tropical rainforest requires substantial rainfall, usually 160 to 300 inches, in comparison to the forty-three inches that New York City receives each year. In severe storms, rain can fall at a rate of eight inches an hour, and in the Amazon basin a river can rise thirty feet overnight. Lying near the equator, the rainforest also gets an abundance of sunshine, with an average annual temperature of 80°F. With the rain and sun, humidity remains high year round, effectively eliminating seasonal changes. These hothouse conditions promote biological activity and growth, which accounts in part for the incredible biodiversity in the rainforest.

In the rainforest everything is interconnected, and nothing is wasted. Some species of army ants, for example, can marshal up to 20 million members to march across the rainforest floor, where they collect everything that has fallen in their path—leaves, twigs, bark, mosses—and take it back to their nests where it is broken down into food for the queen. Other insects flee from this marauding army, only to fly into the waiting mouths of birds and other predators who follow the army ants. The native people of the rainforest use the pincers of live army ants to stitch wounds.

As active and inhabited as the forest floor is, the real life of a rainforest can be found in the trees and vines. The uppermost layer, the canopy, is dominated by the treetops, which rise some 100 to 130 feet. Organisms compete for the sunlight at the top where the brightest flowers flourish. Camouflaged from predators by the colorful tree-top flowers, toucans, hornbills, parakeets, and birds of paradise can afford to show off their flamboyant colors. Most of the wildlife, including spider monkeys, chimpanzees, sloths, marsupials, and squirrels, exists in the trees.

The Amazon rainforest functions as a self-contained hydrological cycle. Water vapor from the Atlantic Ocean falls on the forests, and trans-evaporation releases about half of that rain back into the atmosphere, which creates the region's moist climate. The rainwater that penetrates the canopy provides sustenance, trickling down trees and vines, absorbing compounds from plants and excrement from animals, and picking up important nutrients such as nitrogen, phosphorus, and potassium. Along the way, it feeds the epiphytes, such as mosses, ferns, orchids, fungus, and lichens that grow on other plants and trees. The epiphytes have leaves that form reservoirs for storing rainwater and that also serve as homes for frogs and other animals. The frogs are particularly happy denizens of the rainforest. They absorb water through their skins, and the humidity is ideal for them. Countless varieties of birds, snakes, termites, and spiders also live among the trees and vines, dispersing the seeds of the plants on which they depend.

The Amazon rainforest has also supported indigenous groups for close to ten thousand years. Often, they engage in shifting cultivation, clearing a small patch of land for dwellings and subsistence farming, mainly of maize, beans, and manioc. Small plots of trees are burned, and the ash is used to supply carbon and other nutrients to the poor soil in the rainforest. When, after several years, the soil no longer supports crops, the community moves on. Twenty-five to fifty years later, the land returns to productive use, and the Indians return. The forest supplies trees, vines, and leaves for dwellings, palm trees and bamboo for bows and arrows, vines for baskets and fiber for ropes, and plants for medicines, magical potions, aphrodisiacs, contraceptives, and body decoration. Hunting, fishing, and gathering nuts, fruits, and mushrooms supply other sources of protein and nutrients.

Rubber tappers are another group of people who live from the rainforest of Brazil without destroying it. In the nineteenth and through much of the last century, impoverished Brazilians from the overcrowded, drought-stricken northeast of the country were induced to migrate to the Amazon rainforest to work as rubber tappers. They extract latex by hand from the rubber trees, which are spread out through the forest to protect them from parasites that attack only if the trees are in close proximity. Though the rubber tappers were for many years exploited by intermediary brokers and large American syndicates, foreign control of the Brazilian rubber trade was eventually defeated.

Despite such amazing diversity, the rainforest has always been fragile. Thousands of species of trees, plants, insects, and animals in the rainforests exist only in a very narrow locale. One can find hundreds of specimens of a plant within a small plot of the Amazon rainforest, and yet that species does not exist a short distance away, or anywhere else in the rainforest. Destroy a small plot of the Amazon, and an entire species is destroyed.

In 1800, there were 7.1 billion acres of tropical rainforest throughout the world. By 2000 there were only 3.5 billion acres, with about one-third of those acres in Brazil. The world continues to lose rainforest at the rate of 35 to 50 million acres each year. Over 50 percent of our planet's species

Large tracts of the rainforest continue to be destroyed for soy farming and other agricultural products. Nutrients in the poor rainforest soil are often depleted within ten years.

Credit: © Daniel Beltra/Greenpeace, courtesy of Greenpeace

live in tropical rainforests, including some 5 million species of plants, animals, and insects in Brazil's Amazon basin. It is estimated that one hundred species are lost each day as a result of the destruction of the rainforests. About one quarter of all medicines are derived from plants, not synthetic compounds, and 90 percent of the plants critical to medicine are found only in rainforests. Lost in rainforest destruction are plants that could provide critical medicines, such as those already known to produce curare (used as a muscle sedative in surgery), diosgenin (for birth control pills and to treat arthritis and asthma), and quinine (to treat malaria and pneumonia). Vincristine and vinblastine are derived from rosy periwinkle; they are used not only for the treatment of Hodgkin's disease but have increased the chances of recovery from childhood leukemia from 20 percent to 90 percent.

Besides the plants, trees, and animal species, the world is losing the indigenous peoples of the rainforests. As a result of contact with outsiders, indigenous groups have been decimated by several diseases, including measles, influenza, malaria, venereal disease, tuberculosis, and typhoid, for which they have no developed immunity. Violent conflict, which pitted bows and arrows against shotguns, has also claimed the lives of many natives. Some 6 to 10 million indigenous people lived in the Brazilian Amazon in 1500. In 1900, it was 1 million. Today, fewer than 250,000 people remain.

Until the mid-twentieth century, few substantial threats to the Brazilian forests existed. That changed with the military coup in Brazil in 1964, which led to the economic development of the Amazon and far greater territorial settlement. Rich mineral deposits, land for farming and cattle grazing, and the hydroelectric power potential of rivers were all exploited by the military regime. By settling in the Amazon, Brazilians also secured territory for the state.

While subsequent civilian governments have also been reluctant to put limits on economic development, other pressures contributed to the exploitation of the rainforest resources. Brazil's population grew rapidly in the late twentieth century, and the Amazon, which constitutes half of Brazil's total territory, served to relieve the pressure in the cities. As a result of migration to the Amazon, the non-indigenous population increased from 2 million in the 1960s to 20 million. Inflation grew just as rapidly, becoming as high as 100 percent in the 1970s and 1980s. Not only did land became an attractive, safe investment, but repaying international debt required exporting Brazilian resources to pay the interest with foreign money.

Cattle ranching and soy farms result in the further destruction of the rainforest. After the ranchers destroy large tracts of Brazilian rainforest for cattle grazing, displacing small, subsistence farmers who are forced to move further into the rainforest, the nutrients in the soil are depleted in less than ten years. Once the soil is depleted, the ranchers move, burning more forest and displacing small farmers, natives, and tappers. Soy farms also compete with cattle ranches for deforested land.

The harvesting of tropical hardwood for high-quality furniture in American, Japanese, and Chinese markets not only destroyed the hardwood growth, but also the rubber trees and Brazil nut trees that stood in the way. With the soil exposed to intense heat and heavy rains, it is washed into streams and rivers that become filled with sediment, ruining the surface waters for fishing.

While there have always been lone gold miners exploring the rainforests, government backing led to large-scale mining of gold and iron ore in the mid-1960s. In the 1980s, gold was discovered at Serra Pelada, and the ensuing rush created wealth for some lucky prospectors, but also displaced or killed the natives who happened to live on the land where the gold was discovered. Those who survived continue to pay a price. The gold mining uses mercury to process the ore, and as a result the level of mercury in fish in nearby rivers has increased to four times the level allowed for safe consumption under Brazilian law. The mercury levels in Kayapó children who live near the gold mines are twice as high as the acceptable standard. The environmental disaster in Minamata, Japan, demonstrated how awful the consequences of mercury poisoning can be. In fact, staff from the National Institute for Minamata Disease have helped to monitor the pollution and to assess the health effects of mercury poisoning in the Brazilian rainforest.

In order to supply power for its projects, the government embarked on a drive to construct dams for hydroelectric power. In some cases, developers were in such a rush to flood the land that they did not even bother to harvest the trees. The ensuing soil erosion resulted in a build-up of silt in the reservoirs and feeder streams that made some dams practically useless within a short time after construction. Dam construction also wiped out large tracts of land that were home to indigenous people.

Most of these projects were made possible by government tax subsidies and international financing. Each project required the construction of new roads through the rainforest. And each road led to more soil erosion and more incursions by ranchers and miners. When the cattle ranchers, loggers, and miners continued to exploit resources farther and farther into the rainforest, sometimes resorting to violence and murder, the natives and rubber tappers began to fight back.

Though that resistance involved violence at first, in the 1970s rubber tappers began to unionize and then aggressively resist development through *empates*, or standoffs. When word went out that a rancher was starting to clear an area that the tappers used, large groups of tappers, their wives, and children would gather at the site. They would block the deforestation physically and urge workers to desist from destroying their livelihood. A particularly successful union organizer was a man named Chico Mendes. Mendes, part of a family of rubber tappers, helped tappers gain control over the price and distribution of the rubber they harvested through grassroots organization. He was patient and stubborn, and he traveled long distances by river and through forests to spend time with other rubber tappers, listening to their problems and gaining their trust.

In 1985, with the assistance of the anthropologist Mary Helen Allegretti, and her organization, the Institute for Amazon Studies, Mendes helped to organize the first national congress of rubber tappers. Mendes related the struggle of the rubber tappers to the wider fight for protection of the environment by aligning the movement with international environmental organizations, such as the Natural Resources Defense Council, and the World Wildlife Fund. Environmentalists, too, learned that the fragile rainforest was inhabited by people who depended on it for their livelihood. In 1987 Mendes was invited to the United States to lobby international banking interests against certain development projects in Brazil.

As the peaceful *empate* movement grew, the ranchers decided to stop the rubber trappers by killing the main leadership. Some believed that Mendes's fame would protect him from the violence, but the international attention Mendes received merely intensified the anger of the ranchers. Gunmen were hired to kill hundreds of rubber tappers, as well as activist priests and lawyers. Mendes remained the subject of death threats and was assigned bodyguards by the authorities. On December 22, 1988, while his family and bodyguards were inside his house, Mendes stepped outside to wash and was fatally struck in the chest by a shotgun blast. A local rancher and his son were eventually convicted of the murder, and an investigation continues into the role that other prominent ranchers may have played in the murder. Such murders continue today. In February 2005, Dorothy Stang, an American nun and environmentalist who worked to protect the rainforest and its people from exploitation, was shot dead in the Amazon rainforest after receiving death threats from loggers and landowners. The rancher who ordered and paid for the murder of Sister Dorothy was convicted in the spring of 2007.

The natives of the rainforest also recognized the need to organize, and in 1980 the Indigenous Peoples' Union was formed. The union lobbied for indigenous rights, disseminated information about the natives' way of life, and networked with other groups, including the rubber tappers. At a constitutional convention following the return of a civilian government in Brazil, the union secured rights to traditional lands and the power to influence development that affects their way of life. Like Chico Mendes, the natives received considerable attention from the international environmental community, as well as from the rock star Sting, who has offered substantial support for the Kayapó. Several Kayapó chiefs even went to Washington at the invitation of environmentalists to speak to politicians and the World Bank in opposition to a major Brazilian dam project. When the project was delayed, in part owing to concerns over its environmental impact, the Brazilian government responded by charging the chiefs with sedition. The case brought further international attention, however, forcing the government to dismiss the charges.

Support continues for indigenous groups, rubber tappers, and Brazilian environmentalists in their struggle to stop rainforest destruction. The World Bank and other regional development banks have come under pressure to ensure that economic development projects address the concerns of the natives and the rubber tappers. Some international companies, including cosmetic manufacturers, are working with indigenous groups to make products with tropical plants that are collected without harming the forest. National parks are being established that will set aside large sections of the rainforest.

Reserves have also been established to protect the traditional economic activities of the natives and the rubber tappers and to preserve their culture. The reserves are a form of collective action, in which the local population retains the right to use the land for rubber extraction, nut gathering, or other productive, sustainable uses.

The World Wildlife Fund and the Nature Conservancy, among other organizations, are also engaging in debt-for-nature swaps. A group purchases debt owed to a bank by a developing country and then exchanges that debt for environmental projects in the developing country, such as the purchase of rainforest land, the creation of a park, or the construction of a sewage system. To help encourage consumers to make responsible decisions, the Forest Stewardship Council is working with indigenous groups, industry, and environmental organizations to develop and administer a certification program to support the marketing of timber from forests that have been responsibly managed.

Despite attempts to stop or slow deforestation, there is little enforcement to protect the extractive reserves and demarcated lands from being invaded by ranching and mining interests. Forces within the Brazilian government resist debt-for-nature swaps as fresh attempts by outsiders to seize control of Brazil's natural resources. There are also questions about the economic viability of small-scale extractive activities, in particular the collecting of rubber and Brazilian nuts. Meanwhile, the rainforest continues to be destroyed at an alarming rate. Between 2000 and 2005, more than 51,000 square miles of Brazil's rainforest—an area larger than Greece—was destroyed, and in 2006 and 2007 an additional 9,266 square miles was lost.

This deforestation is disturbing not only because of its impact on the environment of Brazil, but also because of the far-reaching consequences for regional and global climate change. The Amazon rainforest has been described as the "Lungs of our Planet" because it continuously recycles carbon dioxide into oxygen. More than 20 percent of the world's oxygen is produced in the Amazon rainforest. As a result of the burning of the rainforest, an alarming amount of carbon dioxide is released into the atmosphere, and there is no forest to recycle the carbon dioxide. The prevalence of carbon dioxide in the atmosphere is a major contributor to global warming, the next and most pressing environmental disaster facing us today.

GLOBAL CLIMATE CHANGE

Of all the environmental disasters described in this book, Chernobyl had the most far-reaching impact. Yet it pales in comparison to the vast consequences of global climate change. The consequences may not rise to the level of an apocalypse, but they will be disastrous. Just how disastrous will depend largely on what we do right now. Gases such as carbon dioxide and methane are pollutants emitted from various industrial, transportation, and agricultural operations. Once in the air, they trap rays of heat emanating from the earth that otherwise would have discharged into the outer atmosphere. The gases act much like glass in a greenhouse, which has led to their being described widely as greenhouse gases (GHG). While the problem is popularly referred to as global warming, it is more accurate to describe it as global climate change resulting from the emission of GHG. Under certain circumstances and in certain places, the continuing emission of these chemicals will turn some mild locales very warm and turn other mild locales very cold.

Climate change accelerated in the twentieth century, but its most dire effects are just beginning to unfold, and the worst is yet to come. Up to one quarter of all plant and animal species may be wiped out by 2050 as warming makes certain habitats unlivable.

Temperature changes of just several degrees will intensify weather events, including droughts, floods, and hurricanes. Shrinking glaciers in the Andes will no longer provide sufficient water for drinking, irrigation, and hydropower in Bolivia and Peru. Future wars will be fought over water, not oil. If global climate change continues unabated, the warm air currents of the Gulf Stream could be disrupted. Cold, salty water in the North Atlantic sinks, and warm water from the south is pulled north to replace that sinking cold water, moderating what would otherwise be much colder weather. If fresh water from thawing glaciers and increased rainfall disturbs the current balance, the North Atlantic could turn very cold. In such circumstances, Ireland's weather would be comparable to the current weather of Alaska.

The penguins on Robben and Dassen Islands, off South Africa, are threatened by much more than oil spills. Loss of food supplies and other habitat changes resulting from global climate change are already part of the reason that the African penguin and nine other penguin species are being considered for protection under the Endangered Species Act by the U.S. Fish and Wildlife Service. Sea rise causes storm surges that will devastate low-lying areas, including the breeding grounds of the African penguins. Several thousand miles north, those same rising waters threaten the Netherlands, where two-thirds of the people live below sea level. Whole island nations will be wiped off the map of the world, and tens of millions of people will have to be relocated from low-lying coastal areas. These people will become environmental refugees like those forced to flee Chernobyl, Love Canal, Seveso, and Times Beach.

The Chugach Eskimos who live along Alaska's Prince William Sound continue to cope with the lingering effects of the *Exxon Valdez* oil spill, but they face a much greater risk to their subsistence way of life from global warming. Since the 1950s, Alaska's climate has warmed about four degrees, and fish stocks that were once abundant have already declined. Sea ice is farther from shore, thinner, and shorter in duration, making it difficult for walruses and polar bears to hunt. Thawing of the permafrost, which is already destroying highways and utilities, is causing increased ground subsidence, erosion, and landslides. Those who have always lived near the sea, close to their source of life and culture, are finding their villages under direct threat. Alaska's temperature could rise another five degrees by the middle of this century.

The signs of global climate change now appear everywhere, but it was only in the last quarter of the twentieth century that the scientific and policy communities reached a consensus about the threat to the planet. Since some critics continue to deride the significance of global climate change, it is important to understand how that consensus developed, and

A Greenpeace ship moves through thinning sea ice that makes it difficult for walruses to hunt their prey as they depend on sea ice for use as diving platforms, and to travel through feeding areas.

Credit: © Daniel Beltra/Greenpeace, courtesy of Greenpeace

how wide and deep it is. That consensus has to be accepted by everybody before we can develop the political will that is still missing.

Global warming was discovered, oddly enough, by scientists studying ice ages. The glaciations of much of North America and parts of Europe and the thawing that came some ten thousand years ago were cataclysmic events that reconfigured the entire planet. The concern in the scientific community was to understand how the ice age developed, how it ended, and, of most interest, what forces might lead to another ice age descending upon the earth. Various theories were advanced—geologic shifts, volcanic eruptions of ash and cinders into the atmosphere, movement in the oceans, or increases in the levels of carbon dioxide as a result of human action.

These early theories produced several critical insights. First, the change from a non-ice age to an ice age could occur over a relatively short time span, perhaps even hundreds of years. Second, human actions could cause such an earth-changing event. These insights called into question a widely held belief, grounded in part on a theistic view of the world, that the earth and its climate obey orderly, benign processes, that global climate change and other major events are predictable and can occur only over thousands of years.

In the last quarter of the twentieth century, the scientific community began to realize—from studies of fossilized pollen in lake sediments, ice cores, marine fossils, volcanic dust, deep-sea cores, and coral reefs—that it was global warming, not cooling of the earth, that presented the real threat. From a 1965 Boulder, Colorado, conference called "Causes of Climate Change," through an important international conference in Stockholm in the early 1970s, attended by representatives of 113 nations, the scientific community began to acknowledge the possibility of climate warming that might affect the entire planet. These discussions were filled with qualifications that this was only a possibility, something that might happen in the future. By the late 1970s, with studies reported by the U.S. National Academy of Sciences (NAS) and a World Climate Conference in Geneva, Switzerland, the focus crystallized on the increase in CO_2 levels caused by human action, and the language of the discourse strengthened. The studies concluded that increased CO_2 emissions might result in significant changes in global climate and that these changes might be long term.

In the 1980s, scientific research continued, and the consensus grew, at least internationally. In the United States, the Reagan administration resisted these developments. In 1983, the Reagan White House promoted a National Academy of Sciences study that downplayed the possibility of global warming and urged people not to worry. Several scandals involving the U.S. Environmental Protection Agency (EPA), including actions at Times Beach, led to attempts to sanitize the EPA of political influence. As a result, the agency was able to assert some independence in the mid-1980s. In contrast to the NAS, the EPA concluded that catastrophic consequences could be anticipated if global warming continued unabated.

In 1987 the Montreal Protocol on Substances that Deplete the Ozone Layer stipulated that production and consumption of compounds that deplete ozone in the stratosphere—chlorofluorocarbons (CFCs), halons, carbon tetrachloride, and methyl chloroform—were to be phased out. Scientific theory and evidence suggested that, once emitted into the atmosphere, these compounds could significantly deplete the stratospheric ozone layer that shields the planet from damaging UV-B radiation. This international agreement was ratified by the countries responsible for more than 80 percent of world consumption, including the United States. Provisions were included to address the special needs of developing countries with low consumption rates in order to avoid hindering their development. Flexibility in the agreement allowed for adjustments in light of developing scientific evidence without requiring a complete renegotiation. The protocol is a model of collective, cooperative action on the part of countries to protect the global environment, a model that was seen as applicable to global warming. In the late 1980s extremely warm weather was experienced throughout the world, causing loss of life and economic damage and adding urgency to the issue. In 1988 the Intergovernmental Panel on Climate Change (IPCC) was created to assess the potential impact of human-induced climate change and to develop options for adaptation and mitigation. The IPCC was created to periodically gather the current findings of the leading researchers and policymakers on global warming, and to publish the consensus reached in the literature. That consensus in turn serves as the basis for international negotiations and national action to address global warming. Four assessment reports have been issued thus far by the IPCC.

The course of these IPCC reports is instructive. The first, issued in 1990, indicated that the world was warmer but that it was not possible at that time to determine to what extent human action might be contributing to the warming. Whatever the cause, the increasing warmth was a matter of concern. Despite the moderate conclusions, the American administration, under President George H. W. Bush, denounced the report and secretly adopted a policy of stressing any uncertainties as a way of avoiding positive action to address the warming.

The first report played an important role in the discourse at the famous 1992 Rio de Janeiro Conference, known as the Earth Summit, the largest-ever gathering of world leaders. The state of knowledge had advanced enough by 1992 to convince the nations of the world, including the United States, to sign on to the goal of stabilizing GHG emissions at levels that would prevent dangerous human interference with the climate system. However, no specific actions or targets were established.

On the basis of additional studies, the IPCC concluded in its second report in 1995 that "the balance of evidence suggests a discernible human influence on global climate." The consensus in the report was that a doubling of CO_2 levels by the middle of the twenty-first century would cause a 3° to 9°F increase in global average temperatures. The magnitude of this increase had been discussed widely and was not new to scientists in the field. More striking was the growing conviction across the scientific and policy communities that a significant increase in temperature was already occurring.

By the mid-1990s, more divergent voices demanded official recognition of the dangers of global warming and the necessity for corrective action.

Over 2,500 economists, including eight Nobel laureates, endorsed a statement on climate change in February 1997, declaring that the United States could in fact reduce GHG emissions and that "sound economic analysis shows that there are policy options that would slow climate change without harming American living standards, and these measures may in fact improve U.S. productivity in the longer run."[1] British Petroleum broke ranks with the American oil companies and called for action to stop global warming. One of BP's chief executives acknowledged that "companies composed of highly skilled and trained people can't live in denial of mounting evidence gathered by hundreds of the most reputable scientists in the world."[2] Some insurance companies began to take an interest in the issue since they were losing substantial sums of money for weather-related losses; the dire and plausible predictions of even worse global weather catastrophes were getting their attention.

The second assessment report provided key information for the negotiations that led to the adoption of the protocol at a 1997 conference on climate change in Kyoto, Japan. More than 6,000 official delegates attended the conference, along with thousands of representatives of environmental organizations, industries, and the press. To advance the goal set at the Rio Earth Summit, the Kyoto conference was convened to negotiate specific targets and dates by which countries should reduce their GHG emissions. The leading proponents for emission targets were the Europeans, small island nations, and environmental groups; leading the opposition were the oil-producing states, including Australia, Russia, and American oil and auto companies, all united in the Global Climate Coalition, also known as the Carbon Club.

A critical issue that emerged in the Kyoto talks was whether developing countries should be burdened with specific limits on their GHG emissions. Many argued that the developed, industrialized countries had grown wealthy in part by relying on cheap fossil fuel, that global warming was largely the result of emissions from these industrialized countries, and that developing countries should have the same opportunity to strengthen their economies with relatively inexpensive fuel. The Carbon Club seized on this issue and, through advertisements and lobbying, appealed to American nationalism as an excuse for opposing the Kyoto Protocol. It argued that American jobs would be lost to the developing countries unless these countries were also required to share the burden of reducing GHG emissions. Even before the protocol was considered, the U.S. Senate adopted a resolution opposing any agreement that caused substantial economic harm to U.S. interests and that did not require developing countries to adopt targets and dates for compliance.

Delegates to the Kyoto conference determined that it was necessary to establish a timeline for reduction to a specific level of emissions. The targets adopted generally amounted to a modest 5 percent reduction in GHG emissions and were applicable only to certain developed countries and some of the countries with economies in transition, such as those of the former Soviet Union.

The third report by the IPCC in 2001 documented global temperature increases since the 1950s, decreases in snow cover and sea ice, the retreat of mountain glaciers in non-polar regions, a rise in sea levels, increase in global ocean heat, increase in precipitation, increase in frequency of storms, increase in cloud cover, reduction in the frequency of extreme low temperatures, more frequent, persistent, and intense El Niño episodes, and the frequency and intensity of droughts in various regions. This evidence led the scientific and policy community to conclude that "most of the observed warming over the last fifty years is *likely* to have been due to the increase in greenhouse gas concentrations" (emphasis added). The report also estimated that by 2100 the average global temperature would likely rise by 3° to 11°F, an increase from the estimate made in the second report. If no action is taken to reduce emissions, the increase in temperature could be considerably higher. Putting this kind of shift in perspective, the temperature change in the ten thousand years from the last ice age to the preindustrial period was only 10° to 12°F. That change wrought extraordinary environmental shifts, which largely benefited humans. The equivalent change from a warm climate to a very warm climate in less than a hundred years will also produce extraordinary shifts, but these will not be so kind.

The fourth report by the IPCC, published in 2007, concluded that there is a greater than 90 percent chance that climate warming is caused by human beings, and that the projected increase in average global temperature over the next century may be in the range of 5°F. The report also indicated that this warming has caused, and is aggravated by, snow and ice melt, which will lead to a rise in sea levels across the globe. Just how high a rise will occur remains uncertain, but a rise of several feet is likely.

Several aspects of the IPCC reports deserve noting. More than 170 scientists from twenty-five countries contributed to the deliberation and writing of the first report in 1990, and some 200 scientists were involved in a peer review of that report. Four hundred scientists from 26 countries provided expertise for the second report in 1995; 500 scientists from

40 countries provided its peer review. One hundred twenty-two lead authors and 515 other scientists, as well as hundreds of government representatives, participated in the 2001 report; 420 experts reviewed it. The fourth report was prepared by more than 450 lead experts, supported by 800 contributing experts and 2,500 reviewing experts from more than 130 countries. The reports reflect a remarkable consensus among scientists who have studied global climate change for decades.

That consensus was soundly rejected by the George W. Bush administration in the United States. Since the United States is the largest consumer of oil in the world, and the largest emitter of CO_2 and other GHGs, the Bush administration's position was deeply frustrating for leaders of the European Union (EU), and for environmentalists everywhere who adopted the IPCC reports and hoped that implementation of the Kyoto Protocol would finally begin to address climate change. The Bush administration, and some others, called for delay on the grounds that too many uncertainties exist in the scientific data to justify taking corrective action. An editorial in the *New Scientist*, in forcibly countering earlier similar arguments, stated: "Government can't have it both ways. They whine that they can do nothing about global warming because of 'scientific uncertainty,' then cut back on the very science we need to end that uncertainty."[3] Some argue that technological innovation in the future will fix the problem quickly and cheaply, but neither funds nor incentives are provided to motivate or support the development of such solutions. When confronted with the threat that small island nations might be wiped out by sea rise, one opponent of taking action on climate change glibly suggested, "What's wrong with a bit of sea level rise? It is merely changing land use—where there were cows there will be fish."[4] Those whose lives and cultures will be destroyed by even a slight rise in sea level would think otherwise.

Another argument advanced for opposing any action to address climate change is that nothing should be done that might adversely affect economic development. In October 2006, the British government published the *Stern Review on the Economics of Climate Change*, which evaluated the implications of climate change on the world economy. The report concluded that not taking action to ameliorate the effects of climate change would be substantially more costly than taking action now: one percent of global Gross Domestic Product (GDP) must be invested each year in order to mitigate the effects of climate change, whereas doing nothing will cost us up to a 20 percent loss of global GDP.

Times are changing. The administration of George W. Bush has been replaced by the administration of Barack Obama, which is much more favorably committed to addressing the problem. Other countries with a history of inaction or direct opposition to climate change, such as Ireland and Australia, have recently experienced changes in administration and are now moving toward addressing the threat. As the then-British Foreign Secretary Margaret Beckett stated in 2006, "It is now—literally—only one or two fringe scientists and a rather larger number of paid propagandists who still try to deny that climate is changing as a result of human behavior."[5] Hardly a day goes by without a newspaper or television report on yet another piece of evidence that global climate change is having real, immediate, and adverse effects.

In the United States and elsewhere, many local and state governments are taking action. Several states in the northeast have formed a Regional Greenhouse Gas Initiative, which established a cap-and-trade system for reducing GHG from power plants. A cap in GHG emissions is set, then each company either implements improvements at its plants to meet that cap, or purchases GHG credits from other plants that have reduced emissions beyond their requirements. California has also begun to regulate emissions of carbon dioxide from cars, requires electrical utilities to buy energy from only those companies that meet stringent GHG emission standards, requires a 25 percent reduction in GHG emissions by 2020, in part through use of a cap-and-trade system, and has sued automobile manufacturers for contributing to global warming.

Enlightened companies recognize that if they are to operate in a global economy, their European divisions must meet emission targets, and the same technology for achieving these targets can or will be applied in U.S. operations. The U.S. chemical company DuPont, for example, reduced its GHG emissions by 40 percent between 1990 and 2000 throughout its worldwide operations. At the same time it held its energy use constant and increased output by 40 percent. The British oil company BP spent $20 million to reduce its emissions by 20 percent and saved $650 million in production costs.

Many companies see the handwriting on the wall in the United States—some form of control over carbon emissions is inevitable—and they believe that a single form of regulation at the federal level is better than fifty forms of regulation at the state level. The European Union is proceeding with implementation of the Kyoto Protocol targets through a cap-and-trade system to reduce GHG emissions by less than 10 percent.

While such developments are encouraging, they are inadequate. Many countries are reducing GHG emissions only because of the economic downturn rather than making concerted efforts to reduce demand or control emissions. Many countries, including China, are relaxing environmental protections in order to stimulate their troubled economies more quickly. The globalization of the world's economy will continue, and that global economy, including the large developing economies of China and India, will still depend largely on fossil fuel to drive its plants and cars.

Much of the public believes in the reality of global climate change, but people do not necessarily accept that any real change in behavior is required to stop it, or that the situation is a crisis demanding immediate action. Bill McKibben, Jim Hansen, and other environmental activists argue that we may have only a decade to address global climate change before irreversible effects transform our planet forever. As McKibben forewarned recently, the world is slowly awakening to the reality of global warming, "but very few understand with any real depth that a wave large enough to break civilization is forming, and that the only real question is whether we can do anything at all to weaken its force."[6]

Greenhouse gases are inert. They do not dissipate quickly. The greenhouse gases that we discharge into the atmosphere today will be there a hundred years from now. We do not have the luxury of distancing ourselves from this disaster. It is not someone else's problem; it is our own.

The challenge is to convince people to sacrifice now to protect against risks in the distant future. This is a formidable challenge that has to compete with the short-term objectives that dominate corporate bottom lines and the reelection campaigns of politicians.

Individuals must do what they can—drive hybrid or electric cars, switch to energy-efficient appliances and heating systems, waste not—but the larger political realm has to be engaged, on the national and international levels. The problem exists on such a scale that nothing short of direct government intervention will rectify it. What remains is the development of political will. We can vote only for those politicians who are willing and able to face the reality of our own complicity in climate change. We need leaders who are capable of imagining the consequences of global climate change, and who can identify with those who will suffer in the future, including our children, our grandchildren, and their descendants. We need the courage and fortitude, demonstrated by the people in the preceding stories, to take control of events that threaten our environment, and confront those in power who refuse to help us protect it. We are all vulnerable on this borrowed earth.

LIST OF SOME ENVIRONMENTAL ORGANIZATIONS

Greenpeace www.greenpeace.org

Greenpeace is an independent, campaigning organization that uses nonviolent, creative confrontation to expose global environmental problems and promote solutions for a green and peaceful future. Greenpeace's goal is to ensure the ability of the Earth to nurture life in all its diversity. Greenpeace has been campaigning against environmental degradation since 1971 when a small boat of volunteers and journalists sailed into Amchitka, an area north of Alaska where the U.S. government was conducting underground nuclear tests. This tradition of "bearing witness" in a nonviolent manner continues today. For this publication, Greenpeace generously donated a number of photographs illustrating certain of the disasters, as indicated in the credits.

International Fund for Animal Welfare www.ifaw.org

As the world's leading animal welfare organization, IFAW works from its global headquarters in the United States and sixteen country offices to improve the welfare of wild and domestic animals by reducing the commercial exploitation of animals, protecting wildlife habitats, and assisting animals in distress. With projects in more than forty countries, IFAW works both on the ground and in the halls of government to safeguard wild and domestic animals and seeks to motivate the public to prevent cruelty to animals and to promote animal welfare and conservation policies that advance the well-being of both animals and people. IFAW generously donated use of the photograph illustrating the environmental disaster at Robben and Dassen Islands, South Africa, as indicated in the credits.

The Natural Resources Defense Council www.nrdc.org

The Natural Resources Defense Council's purpose is to safeguard the Earth: its people, its plants and animals and the natural systems on which all life depends. NRDC works to restore the integrity of the elements that sustain life—air, land and water—and to defend endangered natural places. It seeks to establish sustainability and good stewardship of the Earth as central ethical imperatives of human society. NRDC affirms the integral place of human beings in the environment and strives to protect nature in ways that advance the long-term welfare of present and future generations. NRDC works to foster the fundamental right of all people to have a voice in decisions that affect their environment. It seeks to break down the pattern of disproportionate environmental burdens borne by people of color and others who face social or economic inequities. Ultimately, NRDC strives to help create a new way of life for humankind, one that can be sustained indefinitely without fouling or depleting the resources that support all life on Earth.

Center for Health, Environment & Justice www.chej.org

Earth First! www.EarthFirst.org

Earthwatch Institute www.earthwatch.org

Environmental Defense www.edf.org

Forest Stewardship Council www.fsc.org

Friends of the Earth International www.foei.org

Green Cross International www.greencrossinternational.net

Green Action Japan www.greenaction-japan.org

International Bird Rescue Research Center www.ibrrc.org

International Union for Conservation of Nature www.iucn.org

Izaak Walton League www.iwla.org

National Audubon Society www.audubon.org

National Wildlife Federation www.nwf.org

Nature Conservancy www.nature.org

Rainforest Action Network www.ran.org

Rainforest Alliance www.rainforest-alliance.org

Sierra Club	www.sierraclub.org
Union of Concerned Scientists	www.ucsusa.org
WE ACT for Environmental Justice	www.weact.org
Wilderness Society	www.wilderness.org
Wildlife Conservation Society	www.wcs.org
World Environmental Organization	www.world.org
World Wildlife Fund/World Wide Fund	www.panda.org

NOTES

INTRODUCTION

1. Barry Commoner, "Failure of the Environmental Effort," *Environmental Law Reporter* 18: 10195 (June 1988).
2. At the back of the book are some Web site addresses for various environmental and citizen groups that can provide further information on ways of protecting our environment.
3. On his Web site, www.environmentaldisasters.info, the author has provided additional material on these environmental disasters, called "Postscripts," that trace implications of the stories and offer some further lessons to be learned.

MINAMATA, JAPAN

1. Throughout the chapter, the dollar equivalents given are for the applicable time period.
2. Ishimure changed the names of many of the patients she wrote about.
3. A pseudonym used by Ishimure in her book.
4. At the request of Tomoko's father, Aileen Smith has withdrawn the photo from further publication.

LONDON, ENGLAND

1. Quoted in William Wise, *Killer Smog: The World's Worst Air Pollution Disaster* (New York: Ballantine Books, 1970), 164–165.

SEVESO, ITALY

1. Michael R. Reich, *Toxic Politics: Responding to Chemical Disasters* (Ithaca: Cornell University Press, 1991), 99, quoting from Marcella Ferrara, *Le Donne di Seveso* (Rome: Editeri Riuniti, 1977).
2. P. Lagadec, "From Seveso to Mexico and Bhopal: Learning to Cope with Crises," in *Insuring and Managing Hazardous Risks: From Seveso to Bhopal and Beyond*, ed. Paul R. Kleindorfer and Howard C. Kunreuther (New York: Springer-Verlag, 1987), 15–16; also Michael R. Reich, *Toxic Politics: Responding*

to Chemical Disasters (Ithaca: Cornell University Press, 1991), 106, both sources citing Laura Conti, *Visto da Seveso, l'Evento Straordinario e l'Amministrazione Ordinaria* (Milan: Feltrinelli, 1977), 18.

LOVE CANAL, NEW YORK

1. Deposition testimony of Leonard Bryant, Trial Exhibit 1697, 246–247; trial transcript page 4592, cited in *United States of America and State of New York v. Hooker Chemicals & Plastics Corporation, et al.*, [cited March 17, 1994].
2. Adeline Gordon Levine, *Love Canal: Science, Politics, and People* (Lexington, MA: Lexington Books, 1982), 29.
3. Levine, *Love Canal*, 34.

RHINE RIVER, SWITZERLAND

1. Mary Shelley, *Frankenstein; or, the Modern Prometheus* (1918), quoted in Mark Cioc, *The Rhine: An Eco-Biography, 1815–2000* (Seattle: University of Washington Press, 2002), 145.
2. John Wicks, "Incomprehensible That Danger Was Not Realised," *Financial Times*, November 15, 1986, 8.

PRINCE WILLIAM SOUND, ALASKA

1. Walter Meganack, Sr., "When the Water Died," in *Season of Dead Water*, ed. Helen Frost (Portland, OR: Breitenbush Books, 1990).
2. Jeff Wheelwright, *Degrees of Disaster, Prince William Sound: How Nature Reels and Rebounds* (New York: Simon & Schuster, 1994), 254.
3. Wheelwright, *Degrees of Disaster*, 159.
4. David Lebedoff, *Cleaning Up: The Story behind the Biggest Legal Bonanza of Our Time* (New York: The Free Press, 1997), 306.
5. Jody Seitz and Rita Miraglia, "Chenega Bay," in *An Investigation of the Sociocultural Consequences of Outer Continental Shelf Development in Alaska*, ed. James A. Fall and Charles J. Utermohle (Alaska: Technical Report No. 160 submitted by the Division of Subsistence of the Alaska Department of Fish and Game to the United States Department of the Interior, March 1995), IV-16.
6. Interview with Tom Dragt in Denita Benyshek, "Journey into the Dead Zone," in *Season of Dead Water*, ed. Helen Frost.

GLOBAL CLIMATE CHANGE

1. Jeremy Leggett, *The Carbon War* (New York: Routledge Press, 2001), 261. Also, http://www.rprogress.org/publications/econstatement.html.
2. Darcy Frey, "How Green Is BP?" *New York Times Magazine*, December 8, 2002, 99.

3. Editorial, "Lucky Break: We Can't Rely on Accidental Discoveries for Vital Information about the Planet," *New Scientist*, September 3, 1997, 3.
4. J. R. Spradley, a former prominent member of President George H. W. Bush's Commerce Department, quoted in Jeremy Leggett, *The Carbon War* (New York: Routledge Press, 2001), 119.
5. Remarks by Rt. Hon. Margaret Beckett M.P., Foreign Secretary, at panel on "Climate Security: Risks & Opportunities for the Global Economy," Council on Foreign Relations, New York City, September 21, 2006.
6. Bill McKibben, "How Close to Catastrophe?" *New York Review of Books*, November 16, 2006, 23–25.

SOURCES

MINAMATA

Almeida, Paul, and Linda Brewster Stearns. "Political Opportunities and Local Grassroots Environmental Movements: The Case of Minamata." *Social Problems* 45 (February 1998): 37–60.

Bardsley, Jan. "Japanese Feminism, Nationalism and the Royal Wedding of Summer '93." *Journal of Popular Culture* 31 (Fall 1997): 189–205.

Breton, Mary Joy. *Women Pioneers for the Environment.* Boston: Northeastern University Press, 1988.

Butler, Steven. "A Bay, and People, Safe Again." *U.S. News & World Report,* August 11, 1997, 41.

Cyranoski, David. "Disputed Diagnoses Hamper Claims of Mercury Poisoning." *Nature,* November 2001, 138.

Efron, Sonni. "Victims Not Ready to Close Books on Minamata Saga." *Los Angeles Times,* August 10, 1997, A1.

Fujie, Sakamoto. "A Family Tragedy." Translated by Sugihara Megumi. *AMPO: Japan Asia Quarterly Review* 27, no. 3 (1997): 30–33.

George, Timothy S. *Minamata: Pollution and the Struggle for Democracy in Postwar Japan.* Cambridge: Harvard University Press, 2001.

Harada, Yoshitaka, and Kaneki Noda. "How It Came About the Finding of Methyl Mercury Poisoning in Minamata District." *Congenital Anomalies* 28 (October 1988): S59–S69.

"History of Minamata Disease." *Minamata City.* http://island.qqq.or.jp/minamata.city/english/me_3clf2.htm.

Huddle, Norie, and Michael Reich with Nahum Stiskin. "Tragedy at Minamata." In *Island of Dreams: Environmental Crisis in Japan,* 102–132. Cambridge, MA: Schenkman, 1987.

Hughes, Jim. "The Journalist" *Camera 35* (April 1974): 2.

———. *W. Eugene Smith: Shadow & Substance: The Life and Work of an American Photographer.* New York: McGraw-Hill, 1989.

Iijima, Nobuka. "Social Structures of Pollution Victims." In Ui, Jun, ed. *Industrial Pollution in Japan.* Tokyo: United Nations University Press, 1992, 154–172.

Interviews by the author with Jun Ui, Aileen Smith, Eiko Sugimoto, Moku, and Michiko Ishimure in November 2003 in Tokyo, Kyoto, Minamata, and Kuwamoto.

Ishimure, Michiko. *Story of the Sea of Camellias*. Translated by Livia Monnet. Kyoto: Yamaguchi, 1983.

———. *Paradise in the Sea of Sorrow: Our Minamata Disease*. Translated by Livia Monnet. Kyoto: Yamaguchi, 1990.

———. "Quo Vadis Homo Nipponicus." Translated by Tashiro Yasuko and Frank Baldwin. *The Japan Interpreter* 8 (Autumn 1973): 392–95.

"Japanese Case Settles after 40 Years." *International Commercial Litigation* (July/August 1996): 8.

Kawamoto, Teruo. "A Shameful Retreat." Special Issue on Minamata Disease. *AMPO: Japan Asia Quarterly Review* 27 (1957): 37–38.

Korn, Pearl. "Smith's Place." *Camera 35* (April 1974): 16.

Locher Freiman, Fran, and Neil Schlager. "Mimamata Bay Mercury Poisoning." In *Failed Technology: True Stories of Technological Disasters*. Vol. II, 326–32. Detroit: UXL, 1995.

Maddow, Ben. "The Wounded Angel: An Illustrated Biography." In *Let Truth Be the Prejudice: W. Eugene Smith His Life and Photographs*. New York: Aperture, 1985.

Masazumi, Harada. "Minamata Disease as a Social and Medical Problem." *Japan Quarterly* XXV (January-March 1978): 20–34.

Masazumi, Harada, with Aileen M. Smith. "Minamata Disease: A Medical Report." In Smith and Smith, *Minamata*, 180–192.

McKean, Margaret A. *Environmental Protest and Citizen Politics in Japan*. Berkeley and Los Angeles: University of California Press, 1981.

"Minamata Disease: The History and Measures," http://www.env.go.jp/en/chemi/hs/minamata2002x.html

Mishima, Akio. *Bitter Sea: The Human Cost of Minamata Disease*. Translated by Richard L. Gage and Susan B. Murata. Tokyo: Kosei, 1992. Originally published in 1977 in Japanese.

Mizoguchi, Kozo. "Japan's Top Court Orders Government to Pay Minamata Mercury Poisoning Victims 22 Years After Case Was Filed." *Associated Press*, October 19, 2004.

Pace, Eric. "Teruo Kawamoto, Victim's Advocate in Mercury Outbreak." *New York Times*, February 22, 1999, B8.

Pierce, Bill. "Homage to a Prickly Pear." *Camera 35* (April 1974): 16.

Pollack, Andrew. "Japan Calls Mercury-Poisoned Bay Safe Now." *New York Times*, July 30, 1997, A9.

———. "Mercury, Mostly Gone from Bay in Japan, Still Poisons Town's Life." *New York Times*, August 23, 1997, 1, 6.

Ross, Catrien. "Minimata Disease Redress Settled." *The Lancet*, December 23/30, 1995, 1695–1696.

Smith, Aileen M. "Why Minamata?" In *Minamata*, by W. Eugene Smith and Aileen M. Smith. Tucson, AZ: Center for Creative Photography, 1981.

Smith, W. Eugene, and Aileen M. Smith. "Minamata, Japan: Life—Sacred and Propane—A Photographic Essay on the Tragedy of Pollution, and the Burden of Courage." *Camera 35* (April 1974): 26–51.

———. *Minamata*. New York: Holt, Rinehart and Winston, 1975.

"Supreme Court Holds State Responsible For Minamata Outbreak." *Asia Pacific Biotech News* 8 (November 15, 2004): 1177.

Swinbanks, David. "Japan Pledges New Aid to Minimata Victims." *Nature*, June 29, 1995, 711.

Thurston, Donald R. "Aftermath in Minamata." *The Japan Interpreter* 9 (Spring 1974): 25–42.

"Top Court Holds State to Account for Minamata." *The Japan Times*, October 16, 2004.

Tremblay, Jean-Francois. "Chisso Settles Most Minamata Disease Cases." *Chemical & Engineering News* 74 (June 3, 1996): 8–9.

Ui, Jun. "Minamata Disease." In Ui, *Industrial Pollution in Japan*, 103–132. Tokyo.

———. "Minamata Disease and Japan's Development." Translated by Sugihara Megumi. *AMPO: Japan Asia Quarterly Review* 27, no. 3 (1997): 18–25.

Ui, Jun, ed. *Industrial Pollution in Japan*. Tokyo: United Nations University Press, 1992.

Upham, Frank K. "Litigation and Moral Consciousness in Japan: An Interpretive Analysis of Four Japanese Pollution Suits." *Law & Society* 10 (1976): 578–619.

Watts, Jonathan. "Minamata Bay Finally Declared Free of Mercury." *The Lancet*, August 9, 1997, 422.

———. "Mercury Poisoning Victims Could Increase by 20,000." *The Lancet*, October 20, 2001, 1349.

Yasumori, Nishi. "Despite My Convulsions, I Haven't Applied." *AMPO: Japan Asia Quarterly Review* 27, no. 3 (1997): 39.

Yoichi, Tani. "The 'Final Settlement': Have We Been Told the Whole Truth." *AMPO: Japan Asia Quarterly Review* 27, no. 3 (1997): 26–29.

LONDON

Abercrombie, G. F. "December Fog in London and the Emergency Bed Source." *The Lancet*, January 31, 1953, 234–235.

Anderson, H. Ross. "Health Effects of Air Pollution Episodes." In Holgate, *Air Pollution and Health*, 461–482.

Ashby, Eric, and Mary Anderson. "Studies in the Politics of Environmental Protection: The Historical Roots of the British Clean Air Act, 1956: III. The Ripening of Public Opinion, 1898–1952." *Interdisciplinary Science Reviews* 2 (September 1977): 190–206.

———. *The Politics of Clean Air.* Oxford: Clarendon, 1981.

Bach, Wilfrid. *Atmospheric Pollution.* New York: McGraw-Hill, 1972. See Chapter 3, "Health Effects of Air Pollution," 43–48.

Ball, D. J., and R. Hume. "The Relative Importance of Vehicular and Domestic Emissions of Dark Smoke in Greater London in the Mid-1970s, the Significance of Smoke Shade Measurements, and an Explanation of the Relationship of Smoke Shade to Gravimetric Measurements of Particulate." *Atmospheric Environment* 11 (1977): 1065–1073.

Bell, Michelle L., and Devra Lee Davis. "Reassessment of the Lethal London Fog of 1952: Novel Indicators of Acute and Chronic Consequences of Acute Exposure to Air Pollution." *Environmental Health Perspectives* 109, supplement 3 (June 2001): 389–394.

Brimblecombe, Peter. *Air: Composition and Chemistry.* Cambridge: Cambridge University Press, 1986.

———. "Air Pollution and Health History." In Holgate, Stephen T., ed. *Air Pollution and Health.* San Diego: Academic Press, 1999. 5–18.

———. *The Big Smoke.* New York: Methuen, 1987.

Brimblecombe, Peter, and Catherine Bowler. "The History of Air Pollution in York, England." *J. Air Waste Management Association* 42 (December 1992): 1562–1566.

Burgess, S. G., and C. W. Shaddick. "Bronchitis and Air Pollution." *Journal of Royal Society of Health* 79 (1959): 10–25.

Davis, Devra. *When Smoke Ran Like Water: Tales of Environmental Deception and the Battle against Pollution.* New York: Basic Books, 2002.

Davis, Devra L., Michelle L. Bell, and Tony Fletcher. "A Look Back at the London Smog of 1952 and the Half Century Since." *Environmental Health Perspectives* 110 (December 2002).

"Design & Equipment in Post-War Housing in Relation to Smoke Prevention." *Smokeless Air* 13 (Autumn 1942): 36–41.

Douglas, C. K. M., and K. H. Stewart. "London Fog of December 5–8, 1952." *Meteorological Magazine* 82 (1953): 67–71.

Eggleston, Simon, Michele P. Hackman, Catherine A. Heyes, James G. Irwin, Roger J. Timmis, and Martin L. Williams. "Trends in Urban Air Pollution in the United Kingdom during Recent Decades." *Atmospheric Environment* 26B (1992): 227–239.

Elsom, Derek M. "Atmospheric Pollution Trends in the United Kingdom" In *The State of Humanity,* edited by Julian L. Simon, 476–490. Cambridge, MA: Blackwell, 1995.

———. *Smog Alert: Managing Urban Air Quality.* London: Earthscan, 1996.

Fry, John. "Effects of a Severe Fog on a General Population." *The Lancet,* January 31, 1953, 235–236.

Great Britain Ministry of Health. *Mortality and Morbidity During the London Fog of December 1952.* London: Her Majesty's Stationery Office, 1954.

Heimann, Harry. "Effects of Air Pollution on Human Health." In *Air Pollution.* Geneva: World Health Organization, 1961. Monograph Series No. 46.

Holgate, Stephen T., ed. *Air Pollution and Health.* San Diego: Academic Press, 1999.

Logan, W. P. "Fog and Mortality." *The Lancet,* January 8, 1949, 78.

———. "Mortality in the London Fog Incident, 1952." *The Lancet,* February 14, 1953, 336–337.

London Calling: The Overseas Journal of the British Broadcasting Corporation, no. 683 (December 4, 1952): 18.

London County Council. *Smoke Nuisance in London: Report of the Chief Officer of the Public Control Department.* London: Public Control Committee, June 1904.

Nemery, Benoit, Peter H. M. Hoet, and Abderrahim Nemmar. "The Meuse Valley Fog of 1930: An Air Pollution Disaster." *The Lancet,* March 3, 2001, 704–708.

Pope, C. Arden III, Richard T. Burnett, Michael J. Thun, Eugenia E. Calle, Daniel Krewski, Kazuhiko Ito, and George D. Thurston. "Lung Cancer, Cardiopulmonary Mortality, and Long-term Exposure to Fine Particulate Air Pollution." *JAMA* 287 (March 9, 2002): 1132–1141.

"Profile: Today Marks the 50th Anniversary of London's Killer Fog." National Public Radio. December 11, 2002, (transcript).

Russell, Honorary Rollo. *Smoke in Relation to Fogs in London.* London: National Smoke Abatement Institution, 1899.

Sanderson, J. B. "The National Smoke Abatement Society and the Clean Air Act (1956)." *Political Studies* 9 (1961): 236–253.

Shaw, Sir Napier, and John Switzer Owens. *The Smoke Problem of Great Cities.* London: Constable & Company, 1925.

Stradling, David, and Peter Thorsheim. "The Smoke of Great Cities: British and American Efforts to Control Air Pollution, 1860–1914." *Environmental History* 4 (January 1999): 6–31.

Times (London), series of articles written December 1–5, 1952.

Times (London), December 6, 1952, 3, 6.

Times (London), December 7, 1952, 8.

Times (London), December 8, 1952, 5, 8.

Times (London), December 9, 1952, 3, 8.

Times (London), December 12, 1952, 9.

Times (London), December 15, 1952, 7.

Times (London), December 16, 1952, 9.

Times (London), December 17, 1952, 7, 9.
Times (London), December 19, 1952, 7, 9.
Times (London), December 22, 1952, 7.
Times (London): December 23, 1952, 7.
Wilkens, E. T. "Air Pollution and the London Fog of December, 1952." *Journal of the Royal Sanitary Institute* 74 (January 1954): 1–21.
Wise, William. *Killer Smog: The World's Worst Air Pollution Disaster.* New York: Ballantine, 1970.

WINDSCALE

Arnold, Lorna. *Windscale 1957: Anatomy of a Nuclear Accident*. New York: St. Martin's, 1992.
Bolter, Harold. *Inside Sellafield.* London: Quartet, 1996.
Dickson, David. "Doctored Report Revives Debate on 1957 Mishap." *Science* 239 (February 5, 1988): 556–557.
Harris, John. "Blast From The Past." *The Guardian*, October 8, 2005.
Herbert, Roy. "The Day the Reactor Caught Fire." *New Scientist,* October 14, 1982, 84.
Jenkins, Russell. "Fifty Years on, the Deadly Legacy of Britain's Worst Nuclear Accident." *TimesOnLine*, October 5, 2007.
Morelle, Rebecca. "Windscale Fallout Underestimated." *BBC News*, October 6, 2007.
Pearce, Fred. "Penney's Windscale Thoughts." *New Scientist*, January 7, 1988, 34.
Simons, Paul. "Model Reveals Reach of Deadly Windscale Plume." *TimesOnLine*, October 10, 2007.
Urquhart, John. "Polonium: Windscale's Most Lethal legacy." *New Scientist*, March 3, 1983, 873.
Williams, Gurney III. "Radioactive accidents." *Science Digest*, August 1971, 10–14.
"Windscale fire remembered." *BBC News* [Video and Audio], October 10, 2007.

SEVESO

"A Drug Giant Plagued by Dioxin's Poison." *Business Week*, May 2, 1983, 42–43.
Alpert, Mark. "Where Have All the Boys Gone?" *Scientific American*, July 1998, 22–24.
Bertazzi, Pier Alberto, and Alessandro di Domenico. "Chemical, Environmental, and Health Aspects of the Seveso, Italy, Accident." In *Dioxin and Health*, edited by Arnold Schecter, 587–632, New York: Plenum, 1994.
Cardillo, Paolo, Alberto Girelli, and Giuseppe Ferraiolo. "The Seveso Case and the Safety Problem in the Production of 2,4,5-Trichlorophenol." *Journal of Hazardous Materials* 9 (1984): 221–234.

Coghlan, Andy. "Did Dioxin Cause Rare Cancers at Seveso?" *New Scientist*, September 4, 1993, 6.

Concise Medical Dictionary. 3rd ed. New York: Oxford University Press, 1990.

Davis, Melton S. "Under the Poison Cloud." *New York Times Magazine*, October 10, 1976, 30.

Dickson, David. "The Embarrassing Odyssey of Seveso's Dioxin." *Science*, June 24, 1983, 1362.

Donelli, Massimo. "Seveso, Inflation of Fear." *CHEMTECH* 19 (March 1989): 140–141.

Fuller, John Grant. *The Poison That Fell from the Sky*. New York: Random House, 1977.

Graham, Frank Jr. "How Are We Fixed for Toxic Clouds?" *Audubon* 79 (January 1977): 138.

Hay, Alastair. *The Chemical Scythe: Lessons of 2,4,5-T and Dioxin*. New York: Plenum, 1982.

Hileman, Bette. "Dioxin Toxicity Research Studies Show Cancer, Reproductive Risks." *Chemical & Engineering News* 71 (September 6, 1993): 5–6.

Johnson, Jeff. "International Body Says Dioxin is a Human Carcinogen." *Environmental Science & Technology* 31 (May 1997): 221A.

Kleindorfer, Paul R., and Howard C. Kunreuther, eds. *Insuring and Managing Hazardous Risks: From Seveso to Bhopal and Beyond*. New York: Springer-Verlag, 1987.

Lagadec, P. "From Seveso to Mexico and Bhopal: Learning to Cope with Crises." In Kleindorfer and Kunreuther, *Insuring and Managing Hazardous Risks*, 13–45.

Landi, M. T., L. L. Needham, G. Lucier, P. Mocarelli, P. A. Bertazzi, and N. Caporaso, "Concentrations of Dioxin 20 Years after Seveso" *The Lancet*, June 21, 1997, 1811.

Landi, Maria Teresa, Dario Consonni, Donald G. Patterson, Jr., Larry L. Needham, George Lucier, Paolo Brambilla, Maria Angela Cazzaniga, Paolo Mocarelli, Angela C. Pesatori, Pier Alberto Bertazzi, and Neil E. Caporaso. "2,3,7,8-Tetrachlorodibenzo-p-Dioxin Plasma Levels in Seveso 20 Years after the Accident." *Environmental Health Perspectives* 106 (May 1998): 273–277.

MacDonald, Marci. "Follow-Up: A Trail of Deadly Waste." *Maclean's*, October 10, 1983, 14.

MacKenzie, Debra. "Seveso: The Dioxin Is Burnt." *New Scientist*, July 4, 1985, 25.

McGinty, Lawrence. "The Graveyard on Milan's Doorstep." *New Scientist*, August 19, 1976, 384–385.

Nanda, Ved P. and Bruce C. Bailey. "Export of Hazardous Waste and Hazardous Technology: Challenge for International Environmental Law." *Denver Journal of International Law and Policy* 17 (1988): 155, 161.

Naschi, G. "Engineering Aspects of Severe Accidents, with Reference to the Seveso, Mexico City, and Bhopal Cases." In Kleindorfer and Kunreuther, *Insuring and Managing Hazardous Risks*, 47–59.

Pocchiari, F., V. Silano, and G. Zapponi. "The Seveso Accident and Its Aftermath." In Kleindorfer and Kunreuther, *Insuring and Managing Hazardous Risks*, 60–78.

"Pollution Case Figure Is Shot to Death in Italy." *New York Times*, February 6, 1980, A6.

Raloff, Janet. "1976 Dioxin Accident Leaves Cancer Legacy." *Science News*, September 4, 1993, 149.

———. "Dioxin Cuts the Chance of Fathering a Boy." *Science News*, June 3, 2000, 358.

Renn, Ortwin. "Risk Communication at the Community Level: European Lessons from the Seveso Directive." *JAPCA* 39 (October 1989): 1301–1308.

Schneider, Keith. "Two Decades after Toxic Blast in Italy, Several Cancers Show Rise." *New York Times*, October 26, 1993, C4.

———. "Fetal Harm is Cited as Primary Hazard in Dioxin Exposure." *New York Times*, May 11, 1994, 1.

Signorini, S., P. M. Gerthoux, C. Dassi, M. Cazzaniga, P. Brambilla, N. Vincoli, and P. Mocarelli. "Environmental Exposure to Dioxin: The Seveso Experience." *Andrologia* 32 (2000): 263–270.

Steenland, Kyle, Pier Bertazzi, Andrea Baccarelli, and Manolis Kogevinas, "Dioxin Revisited: Developments since the 1997 IARC Classification of Dioxin as a Human Carcinogen." *Environmental Health Perspectives* 112 (September 2004): 1265–1268.

Stone, Richard. "New Seveso Findings Point to Cancer." *Science*, September 10, 1993, 1383.

Strigini, Paolo. "The Italian Chemical Industry and the Case of Seveso." *UNEP Industry and Environment* 6 (October-December 1983): 16.

"Study Finds Dioxin-Cancer Link." *Chemical Marketing Reporter* 244 (September 13, 1993): 9.

"Town in Italy's Toxic Area Misses Children It Sent Away." *New York Times*, August 17, 1976.

Warner, Marcella, Brenda Eskenazi, Paolo Mocarelli, Pier Mario Gerthoux, Steven Samuels, Larry Needham, Donald Patterson, and Paolo Brambilla, "Serum Dioxin Concentrations and Breast Cancer Risk in the Seveso Women's Health Study." *Environmental Health Perspectives* 110 (July 2002): 625–628.

Webster, Thomas, and Barry Commone. "Overview: The Dioxin Debate." In *Dioxins and Health*, edited by Arnold Schecter, 1, 22–28, New York: Plenum, 1994.

Whiteside, Thomas. *The Pendulum and the Toxic Cloud: The Course of Dioxin Contamination*. New Haven: Yale University Press, 1979.

LOVE CANAL

Many of the facts about the operations of the Hooker company, and other relevant material, are derived from the trial transcripts and other legal submissions in the lawsuit by the United States and the State of New York against the chemical company to recover the over $200 million spent by the governments to investigate and remediate Love Canal, including the buyouts of several hundred homes. Decisions from the court in this case are cited below. The author was one of the trial attorneys for the State of New York in this litigation.

Binns, Jessica. "Remediation: Cleanup Complete at Love Canal." *Civil Engineering News* (December 2004): 22.

Brown, Michael H. *Laying Waste: The Poisoning of America by Toxic Chemicals.* New York: Pantheon, 1980.

DeRosa, Christopher T., and Hugh Hansen. "The Impact of 20 Years of Risk Assessment on Public Health." *Human and Ecological Risk Assessment* 9 (2003): 1219–1228.

Description and Plan of the Model City, Located at Lewiston, Niagara County, NY, Designed to be The Most Perfect City in Existence. Lewiston, NY: Model Town Company, 1893.

Epstein, Samuel, Lester O. Brown, and Carl Pope, *Hazardous Waste in America.* San Francisco: Sierra Club, 1982.

Fletcher, Thomas H. *From Love Canal to Environmental Justice.* Peterborough, Ontario: Broadview, 2003.

Fowlkes, Martha R., and Patricia Y. Miller. *Love Canal: The Social Condition of Disaster.* Washington, DC: Federal Emergency Management Agency, 1982.

Gibbs, Lois Marie. *Love Canal: My Story.* Albany: State University of New York Press, 1982.

Hammer, Armand, with Neil Lyndon. *Hammer.* New York: Putnam Perigee, 1988.

Hooker Electrochemical Company. *Elon Huntington Hooker, 1869–1938: A Tribute to Our Founder.* Niagara Falls, NY: n/d.

Hooker Electrochemical Company after Twenty-Five Years: Manufacturers of High Grade Chemicals. New York: Hooker Electrochemical Co., 1929.

Layzer, Judith A. "Love Canal: Hazardous Waste and the Politics of Fear." In *The Environmental Case: Translating Values into Policy.* Washington, DC: CQ, 2002.

Levine, Adeline Gordon. *Love Canal: Science, Politics, and People.* Lexington, MA: Lexington, 1982.

Mazur, Allan. *A Hazardous Inquiry: The Rashomon Effect at Love Canal.* Cambridge, MA: Harvard University Press, 1998.

New York State Department of Health. *Habitability Decision: Report of Habitability, Love Canal Emergency Declaration Area.* Albany, NY: September 27, 1988.

New York State Department of Health. *Love Canal Follow-up Health Study: Project Report to ATSDR, Public Comment Draft.* Albany, NY: October 2006.

New York State Department of Health. *Supplement to the Love Canal Emergency Declaration Area Proposed Habitability Criteria*, Appendix 6: Love Canal Chronology. Albany, NY: September 1988.

Thomas, Robert E. *Salt and Water, Power and People: A Short History of Hooker Electrochemical Company*. Niagara Falls, NY: Hooker Electrochemical Company, 1955.

Thompson, Carolyn. "Twenty-five Years Later, the Battle over Love Canal Goes On." *Albany Times Union*, August 3, 2003, D4.

U.S. Congress. Office of Technology Assessment. *Habitability of the Love Canal Area: An Analysis of the Technical Basis for the Decision on the Habitability of the Emergency Declaration Area*. Washington, DC: U.S. Government Printing Office, June 1983.

United States v. Hooker Chemicals & Plastics Corporation, 680 F.Supp. 546 (W.D.NY 1988).

United States v. Hooker Chemicals & Plastics Corporation, 722 F.Supp. 960 (W.D.NY 1989).

United States v. Hooker Chemicals & Plastics Corporation, 850 F.Supp. 993 (W.D.NY 1994).

Weinberg, Steve. *Armand Hammer: The Untold Story*. Boston: Little, Brown, 1989.

Worden, Amy. "Twenty-five Years Later, Love Canal's Lessons Still Resonate." *Philadelphia Inquirer*, August 1, 2003, A1, A12.

THREE MILE ISLAND

Davis, Lee. *Encyclopedia of Man-Made Catastrophes*. London: Headline, 1994.

Eisenbud, Merril. *Environmental Radioactivity: From Natural, Industrial, and Military Sources*. 3rd ed. New York: Academic Press, 1989.

Ford, Daniel F. *Three Mile Island: Thirty Minutes to Meltdown*. New York: Penguin, 1982.

Goldsteen Raymond L., and John K. Schorr. *Demanding Democracy after Three Mile Island*. Gainesville: University of Florida Press, 1991.

Gray, Mike, and Ira Rosen. *The Warning: Accident at Three Mile Island*. New York: Norton, 1982.

Hampton, Wilborn. *Meltdown: A Race against Nuclear Disaster at Three Mile Island—A Reporter's Story*. Cambridge: Candlewick, 2001.

Holzman, David C. "Cancer and Three Mile Island: No Significant Increase in Five-Mile Radius." *Environmental Health Perspectives* 111 (March 2003): A166–167.

Hopkins, Andrew. "Was Three Mile Accident a 'Normal Accident'?" *Journal of Contingencies and Crisis Management* 9 (June 2001): 65–72.

Leppzer, Robert. *Voices from Three Mile Island: The People Speak Out*. Trumansburg, NY: Crossing Press, 1980.

Mangano, Joseph. "Three Mile Island: Health Study Meltdown." *Bulletin of the Atomic Scientists* (September/October 2004): 31–35.

Newton, David E.. "Three Mile Island Accident: Middletown, Pennsylvania (1979)." In *When Technology Fails: Significant Technological Disasters, Accidents, and Failures of the Twentieth Century*, edited by Neil Schlager, 510–516. Detroit: Gale Research, 1994.

New York Times, March 31, 1979, A1, A7, A9.

New York Times, April 1, 1979, A1, A29.

New York Times, April 3, 1979, A4, A14.

New York Times, April 4, 1979, A1, A14–17.

Osif, Bonnie A., Anthony J. Baratta and Thomas W. Conkling. *TMI Twenty-Five Years Later: The Three Mile Island Nuclear Power Plant Accident and Its Impact*. University Park: Pennsylvania State University Press, 2004.

Philadelphia Inquirer, March 31, 1979, 1A.

Stephens, Mark. *Three Mile Island*. New York: Random House, 1980.

Talbott, Evelyn O., Ada O. Youk, Kathleen P. McHugh, Jeffrey D. Shire, Aimin Zhang, Brian P. Murphy, and Richard A. Engberg. "Mortality among the Residents of the Three Mile Island Accident Area: 1979–1992." *Environmental Health Perspectives* 108 (June 2000): 545–552.

Talbott, Evelyn O., Ada O. Youk, Kathleen P. McHugh-Pemu, and Jeanne V. Zborowski. "Long-Term Follow of the Residents of the Three Mile Island Accident Area: 1979–1992." *Environmental Health Perspectives* 111 (March 2003): 341–348.

In re TMI Litigation. 193 F.3d 613 (3d Cir. 1999); 199 F.3d 158 (3d Cir. 2000).

Walker, J. Samuel. *Three Mile Island: A Nuclear Crisis in Historical Perspective*. Berkeley: University of California Press, 2004.

Washington Post, March 30, 1979, A2.

Washington Post, March 31, 1979, A1, A9.

Washington Post, April 1, 1979, A1, A6.

Washington Post, April 2, 1979, A1.

Wing, Steve. "Objectivity and Ethics in Environmental Health Science" *Environmental Health Perspectives* 111 (November 2003): 1809–1818.

TIMES BEACH

Carter, Coleman D., Renate D. Kimbrough, John A. Liddle, Richard E. Cline, Mathew M. Zack, Jr., William F. Barthel, Robert E. Koehler, and Arthur E. Phillips. "Tetrachlorodibenzodioxin: An Accidental Poisoning Episode in Horse Arenas." *Science*, May 1975, 738–740.

Freivogel, William, Marjorie Mandel, Jo Mannies, and Lawrence M. O'Rourke. "DIOXIN: Quandry for the '80s. A Comprehensive Survey." *St. Louis Post-Dispatch*, November 13, 1983, supplement.

Goodman, Adam. "Wildlife, Flowers Signal Rebirth of Times Beach as Route 66 State Park." *St. Louis Post-Dispatch*, September 5, 1999.

History of Route 66. http://www.national66.com/66hstry.html.

Kelly, Susan Croce. *Route 66: The Highway and Its People*. With photos by Quinta Scott. Norman: University of Oklahoma Press, 1988.

Reinhold, Robert. "Missouri Dioxin Cleanup: A Decade of Little Action." *New York Times*, February 20, 1983, 1, 54.

"Route 66 Facts and Trivia." *Historic Route 66*. http://www.historic66.com.

Simon, Stephanie. "Park Blossoms over Remnants of Dioxin-stricken Town; Officials Say Site Is Safe 15 Years after Times Beach Was Wiped Off Map." *Milwaukee Journal Sentinel*, October 31, 1999.

St. Louis Post-Dispatch, October 28, 1982, A-1.

St. Louis Post-Dispatch, November 2, 1982, A-1, A-4.

St. Louis Post-Dispatch, November 21, 1982, A-7.

St. Louis Post-Dispatch, November 26, 1982, 17.

St. Louis Post-Dispatch, December 8, 1982, 11A.

St. Louis Post-Dispatch, November 13, 1983, 8.

United States v. NEPACCO et al. 579 F.Supp. 823 (W. D. Mo. 1984) [Denny farm site].

United States v. Russell M. Bliss et al. 667 F.Supp. 1298 (E. D. Mo. 1987) [six sites: Frontenac, Rosati, four horse farms].

Wallis, Michael. *Route 66: The Mother Road*. New York: St. Martin's, 1990.

"Witness Says Driver Lied about Spraying Dioxin." *New York Times*, January 27, 1983, A-9.

BHOPAL

"The Bhopal Legacy: An Interview with Dr. Rosalie Bertell." *Multinational Monitor* 18 (March 1997): 26.

Bhushan, Bharat, and Arun Subramaniam. "Bhopal: What Really Happened?" *Business India* 182 (February 25-March 10, 1985): 102–116.

Cassels, Jamie. "The Uncertain Promise of Law: Lessons from Bhopal." *Osgoode Hall Law Journal* 29 (1991): 1–50.

Daehler, Curtis C., and Shyamal K. Majumdar. "Industrial Disasters: Lessons from Bhopal." In *Natural and Technological Disasters: Causes, Effects and Preventive Measures*, edited by S. K. Majumdar, G. S. Forbes, E. W. Miller, and R. F. Schmalz, 310–321. Easton, PA: The Pennsylvania Academy of Science, 1992.

Dagani, Ron. "Data on MIC's Toxicity Are Scant, Leave Much To Be Learned." *C&EN* 63 (February 11, 1985): 37–40.

David, Luke. "Night of the Gas." *New Internationalist* 352 (December 2002): 34–35.

De Grazia, Alfred. *A Cloud over Bhopal: Causes, Consequences, and Constructive Solutions*. Bombay: Kalos Foundation for the India-America Committee for the Bhopal Victims, 1985.

Dhara, Ramana, and Rosaline Dhara. "The Union Carbide Disaster in Bhopal: A Review of Health Effects." *Archives of Environmental Health* 57 (September/October 2002): 391–404.

Diamond, Stuart. "The Bhopal Disaster: How It Happened." *New York Times*, January 28, 1985, A1, A6.

———. "The Disaster in Bhopal: Workers Recall Horror." *New York Times*, January 30, 1985, A1, A6.

Ember, Lois R. "Technology in India: An Uneasy Balance of Progress and Tradition." *Chemical & Engineering News* 63 (February 11, 1985): 61–65.

Everest, Larry. *Behind the Poison Cloud: Union Carbide's Bhopal Massacre.* Chicago: Banner, 1985.

Greenpeace, "Bhopal Water," report found on Greenpeace website in its archive as archive.greenpeace.org/toxics/documents/Bhopalwater.pdf.

"Has the World Forgotten Bhopal?" *The Lancet*, December 2, 2000, 1863.

Hazarika, Sanjoy. *Bhopal: The Lessons of a Tragedy.* New York: Penguin, 1987.

Hedges, Chris. "A Key Figure Proves Elusive in a U.S. Suit Over Bhopal." *New York Times*, March 5, 2000, international edition, 4.

Huggler, Justin. "Bhopal: A Living Legacy of Corporate Greed." *The Independent*, December 2, 2004.

Jayaraman, Nityanand. "Slow Motion Bhopal." *Multinational Monitor* 18 (April 1997): 6.

Kumar, Sanjay. "Bhopal Disaster Victims' Cases Reopened." *The Lancet*, June 15, 1996, 1687.

MacKenzie, Debra. "Fresh evidence on Bhopal disaster." *New Scientist*, December 7, 2002, 6–7.

McFadden, Robert D. "India Disaster: Chronicle of a Nightmare." *New York Times*, December 10, 1984, A1, A6.

Ng, Delvin. "Call to alleviate long-term effects of Bhopal gas disaster." *The Lancet*, December 14, 1996, 1652.

Orr, David. "Contaminated Lives." *Irish Times*, November 27, 2004, 9.

Reinhold, Robert. "Disaster in Bhopal: Where Does Blame Lie?" *New York Times*, January 31, 1985, A1, A8.

Sharma, Dinesh C. "Bhopal's Health Disaster Continues to Unfold." *The Lancet*, September 14, 2002, 859.

———. "Bhopal: 20 Years On." *The Lancet*, January 8, 2005, 111–112.

Shastri, Lalit. *Bhopal Disaster: An Eye Witness Account.* New Delhi: Criterion, 1985.

Shrivastava, Paul. *Bhopal: Anatomy Of A Crisis.* 2nd ed. London: Paul Chapman, 1992.

"Slum Dwellers Unaware of Danger." *New York Times*, January 31, 1985, A8.

"A Stage Set for a Disaster." *New York Times*, January 30, 1985, A6.

Stevens, William K. "In Bhopal, Signs of Tragedy Are Everywhere." *New York Times*, December 10, 1984, A7.

"Union Carbide toxic vapor leak," In *Failed Technology: True Stories of Technological Disasters*, by Fran Locher Freiman and Neil Schlager. Vol. II, 341–349. Detroit: UXL, 1995.

Union Carbide Web site on the disaster. http://www.bhopal.com.

Wilkins, Lee. *Shared Vulnerability: The Media and the American Perceptions of the Bhopal Disaster.* New York: Greenwood, 1987.

Worthy, Ward. "Methyl Isocyanate: The Chemistry of a Hazard." *Chemical & Engineering News* 63 (February 11, 1985): 27–28.

CHERNOBYL

"Baa, Baa, Blue Sheep, Have You Any Caesium?" *New Scientist*, April 23, 1987, 49.

Baverstock, Keith, and Dillwyn Williams. "The Chernobyl Accident 20 Years On: An Assessment of the Health Consequences and the International Response." *Environmental Health Perspectives* 114 (September 2006): 1312–1317.

Bond, Michael. "Cheating Chernobyl." *New Scientist*, August 21, 2004, 44–47.

Byckau, Mikhail. "Chernobyl: Once and Future Shock, A Liquidator's Story." Translated by Vera Rich.

Cardis, Elisabeth, and Alexey E. Okeanov. "What Is Feasible and Desirable in the Epidemiologic Follow-up of Chernobyl." *The Radiological Consequences of the Chernobyl Accident*, 835–50. Luxembourg: Office for Official Publications of the European Communities, 1996.

Chernobyl Forum. *Chernobyl's Legacy: Health, Environmental, and Socio-Economic Impacts.* Vienna: International Atomic Energy Agency, 2005.

"Chernobyl's Death Toll Could Reach into Thousands." *The Nation's Health*, November 2005, 10.

Chivers, C. J. "New Sight in Chernobyl's Dead Zone: Tourists." *New York Times*, January 15, 2005, A1.

Coghlan, Andy. "Why hot sheep need soft drinks." *New Scientist*, April 11, 1994, 22.

Demidchik, E. P., I. M. Drobyshenskaya, E. D. Cherstvoy, L. N. Astakhova, A. E. Okeanov, T. V. Vorontsova, and M. Germenchuk. "Thyroid Cancer in Children in Belarus," *The Radiological Consequences of the Chernobyl Accident*, 677–682.

Ebel, Robert E. *Chernobyl and Its Aftermath: A Chronology of Events.* Washington, DC: The Center for Strategic and International Studies, 1994.

Environmental and Health Consequences in Japan Due to the Accident at Chernobyl Nuclear Reactor Plant. Chiba, Japan: Natural Institute of Radiological Sciences, 1988.

Goldman, Marvin. "Chernobyl: A Radiological Perspective." *Science*, October 20, 1987, 622–623.

Gould, Peter. *Fire in the Rain: The Democratic Consequences of Chernobyl.* Cambridge, U.K.: Polity, 1990.

Henrich, E. "Chernobyl—Its Impact on Austria." *The Science of the Total Environment* 70 (1988): 433–454.

Herbert, Roy. "Chernobyl: How the Cover Was Blown." *New Scientist*, April 23, 1983, 34.

Hills, David M. "Life in the Hot Zone around Chernobyl." *Nature*, April 25, 1996, 665–666.

Hohenemser, Christoph. "Chernobyl, 10 Years Later." *Environment* 38 (April 1996): 3.

Hohenemser, Christoph, and Ortwin Renn. "Chernobyl's Other Legacy: Shifting Public Perceptions of Nuclear Risk." *Environment* 30 (April 1988): 4–6.

Howard, Brenda, and Francis Livens. "May Sheep Safely Graze?" *New Scientist*, April 23, 1987, 46–49.

International Advisory Committee. *The International Chernobyl Project: An Overview*, 19, 32–35, 42–53. Vienna: International Atomic Energy Agency, 1991.

International Atomic Energy Agency. *One Decade after Chernobyl: Summing Up the Consequences of the Accident.* Vienna: IAEA, 1996. Summary of International Conference held April 8–12, 1996.

"Little to Fear but Fear Itself." *Economist*, September 10, 2005, 77–78.

Lofstedt, Ragnar E., and Allen L. White. "Chernobyl: Four Years Later, the Repercussions Continue." *Environment* 32 (April 1990): 2–5.

MacKenzie, Debra. "The Rad-Dosed reindeer." *New Scientist*, December 18, 1986, 37–40.

Marples, David R. "The Chernobyl Disaster: Its Effect on Belarus and Ukraine." In *The Long Road to Recovery: Community Responses to Industrial Disaster*, edited by James K. Mitchell, 184–229. New York: United Nations University Press, 1996.

Medvedev, Zhores. *The Legacy of Chernobyl.* New York: Norton, 1990.

Motavalli, Jim. "Living With Radiation: Human Health and Nuclear Exposure." *E Magazine* (July/August 2007): 35.

Mould, Richard F. *Chernobyl Record: The Definitive History of the Chernobyl Catastrophe.* Philadelphia: Institute of Physics Pub., 2000.

Mulvey, Stephen. "Chernobyl's Continuing Hazards." *BBC News*, April 26, 2006.

Nuclear Energy Agency. *Chernobyl: Assessment of Radiological and Health Impacts.* Organization for Economic Co-Operation and Development, 2002.

Nussbaum, Rudi H. "The Chernobyl Nuclear Catastrophe: Unacknowledged Health Detriment." *Environmental Health Perspectives* 115 (May 2007): A238-A239 (Correspondence).

Park, Chris C. *Chernobyl: The Long Shadow.* London: Routledge, 1989.

Peterson, Scott. "After Disaster: The People Who Call Chernobyl home." *Christian Science Monitor*, December 15, 2000, 7.

Petridou, E., D. Trichopoulos, N. Dessypris, V. Flytzani, S. Haidas, M. Kalmanti, D. Koliouskas, H. Kosmidis, F. Piperopoulou, and F. Tzortzatou. "Infant Leukaemia after *in utero* Exposure to Radiation from Chernobyl." *Nature*, July 25, 1996, 352–353.

Petryna, Adriana. *Life Exposed: Biological Citizens after Chernobyl*. Princeton, NJ: Princeton University Press, 2002.

"Recalculating the Cost of Chernobyl." *Science*, May 8, 1987, 658–659.

Shcherbak, Iurii. *Chernobyl: A Documentary Story*. Translated by Ian Press. London: Macmillan, 1989.

Shcherbak, Yuri M. "Ten Years of the Chornobyl Era." *Scientific American*, April 1996, 44, 49.

"Shield Is Springing Nuke Leaks, New Cover for Reactor Is…on Drawing Board." *Associated Press*, April 26, 2006.

Stone, Richard. "The Explosion That Shook the World." *Science*, April 19, 1996, 352–354.

Travis, John. "Chernobyl Explosion: Inside Look Confirms the Radiation." *Science*, February 11, 1994, 750.

———. "Inside Look Confirms More Radiation." *Science*, February 11, 1994, 750.

United Nations. *Chernobyl's Legacy: Health, Environmental and Socio-Economic Impacts*. Report of the United Nations Chernobyl Forum, August 2005.

Yaroshinskaya, Alla. *Chernobyl: The Forbidden Truth*. Translated by Michele Kahn and Julia Sallabank. Lincoln: University of Nebraska Press, 1995.

The Watt Committee on Energy. *Five Years After Chernobyl: 1986–1991, A Review*. London: Watt Committee, 1991.

Webb, Jeremy. "Thyroid Cancer Takes Its Toll on Chernobyl's Children." *New Scientist*, April 1, 1995, 7.

"Who Pays the Bill for Radioactive Pollution?" *New Scientist*, April 23, 1987, 46.

Williams, Dillwyn. "Cancer after Nuclear Fallout: Lessons from the Chernobyl Accident." *Nature Reviews* 2 (July 2002): 543–549.

Williams, Nigel, and Michael Balter. "Chernobyl Research Becomes International Growth Industry." *Science*, April 19, 1996, 355.

Wynne, Bran. "Sheepfarming after Chernobyl: A Case Study in Communicating Scientific Information." *Environment* 31 (March 1989): 14.

RHINE RIVER

Anderberg, S., and W. M. Stigliani. "An Integrated Approach for Identifying Sources of Pollution: The Example of Cadmium Pollution in the Rhine River Basin." *Water Science and Technology* 29 (1994): 61–67.

Ardill, John. "British Firm to Burn 20 Tonnes of Contaminated Swiss Waste." *The Guardian*, June 7, 1988.

Bernauer, Thomas. "Protecting the Rhine River against Chloride Pollution." In *Institutions for Environmental Aid*, edited by Robert O. Keohane and Marc A. Levy, 201–232. Cambridge, MA: MIT Press, 1996.

Bernstein, Richard. "Letter from Germany: No Longer Europe's Sewer, but Not the Rhine of Yore." *New York Times*, April 21, 2006, A4.

Betts, Paul. "Sandoz Pays for Rhine Spill." *Financial Times*, September 30, 1987, Section I, 2.

Chichester, Page. "Resurrection on the Rhine." *International Wildlife Magazine*, September-October 1997.

Cioc, Mark. *The Rhine: An Eco-Biography, 1815–2000*. Seattle: University of Washington Press, 2002.

Curlee, Lilian. "Steps toward Reintroduction of Natural Systems into the Management of the Rhine River." *Restoration and Reclamation Review* 4 (Spring 1999). Student on-line journal, Department of Horticultural Science, University of Minnesota.

Davies, Gareth Huw. "Slow Flow of Warnings on River of Death." *Times* (London), November 16, 1986.

Dawkins, William. "Swiss Face Criticism over Sandoz Accident." *Financial Times*, November 12, 1986, 3.

Dieperink, Carel. "International Regime Development: Lessons from the Rhine Catchment Area." *TDRI Quarterly Review* 12 (September 1997): 27–35.

Drozdiak, William. "Cleanup Efforts Bring Fish Back to Rhine; for 1st Time in Decades, Species Reach Upper River." *Dallas Morning News*, March 31, 1996.

Dullforce, Willam. "Sandoz Warehouse Met Swiss Safety Rules." *Financial Times*, November 11, 1986.

England, John. "Second Swiss Company Admits Dumping Poison into Rhine before Blaze." *Times* (London), November 12, 1986.

Engler, Jerry. "Pesticides: Atrazine Use Down, Ag Experts Report." *Greenwire*, December 17, 2004.

"Europeans Do It to Each Other." *The Economist*, November 15, 1986, 43.

"Europe's Cities; Basle; Hands across the Borders." *The Economist*, November 8, 1986, 61.

"Greater Rhine Pollution Charged." *Washington Post*, November 15, 1986, A29.

Glass, A. and C. Snyder. "Shocked into Action." *Harvard International Review* 18, Issue 4 (Fall 1996): 48–52.

Hayes, Tyrone, Kelly Haston, Mable Tsui, Anhthu Hoang, Cathryn Haeffele, Aaron Vonk. "Herbicides: Feminization of male frogs in the wild." *Nature* 419 (October 31, 2002): 895–896.

Hull, Jennifer B. "A Proud River Runs Red." *Time* Magazine, November 24, 1986, 36.

Kiss, Alexandre. "The Protection of the Rhine Against Pollution." *Natural Resources Journal* 25 (July 1985): 613–37.

Kurlansky, Mark J. "Who is Killing the Rhine?" *Environment* 24 (September 1982): 41.

Lenz, Sara. "Basle to Test 600 Feared Affected by Chemical Fire." *The Guardian*, November 20, 1986.

———. "Firm Polluted River with Weed Killer 'for a Year.'" *The Guardian*, November 15, 1986.

———. "New Chemical Accident Leaves the Swiss Coughing." *The Guardian*, November 21, 1986.

———. "Second Firm Admits Rhine Poison Spill; Swiss Chemical Company Ciba-Geigy Dumps Toxic Compounds in River." *The Guardian*, November 12, 1986.

Lenz, Sara, and Derek Brown. "Rhine Could Face Another Pollution Disaster; Threat from Toxic Sludge Still Lying on River Bed." *The Guardian*, November 14, 1986.

Lewis, Paul. "Europe Mired in Bickering Over Who Dumps What." *New York Times*, November 16, 1986, Section 4, 3.

———. "Huge Chemical Spill in the Rhine Creates Havoc in Four Countries." *New York Times*, November 11, 1986, A1.

Malle, Karl-Geert. "Cleaning Up the Rhine River." *Scientific American*, January 1996, 70–75.

Markham, James M. "Rhine Pollution Is Tricky Issue in West Germany." *New York Times*, December 21, 1986, Section 1, 18.

Marsh, David. "Fishermen to Claim for Loss of Eel Stocks Worth 35.5 M Pounds." *Financial Times*, November 13, 1986, Section I, 2.

———. "W. Germans Worry Over Long Term Harm to Rhine." *Financial Times*, November 11, 1986, 3.

McCartney, Robert J. "Europe Tries to Cope with a Poisoned Rhine." *Washington Post*, November 14, 1986, A29.

———. "Rhine Toxic Spill Draws Protests; Swiss, Company Are Criticized as Slow in Disclosing Mishap." *Washington Post*, November 12, 1986, A21.

Montgomery, Paul L. "These Days, Watch on the Rhine Reveals New Perils of Pollution." *Los Angeles Times*, November 30, 1986, Part 5, 2.

Netter, Thomas, "Anger Along the Rhine Grows after Chemical Spill." *New York Times*, November 12, 1986, A8.

———. "Mercury a Key Concern in Rhine Spill." *New York Times*, November 15, 1986, Section 1, 3.

———. "New Chemical Accident Sends a Cloud Over Basel." *New York Times*, November 21, 1986, A13.

———. "Poisons Pumped from Rhine." *Chicago Tribune*, November 18, 1986, Section C, 10.

———. "Rhine Disaster Brings Tide of Anger to Swiss." *Chicago Tribune*, November 12, 1986, Section C, 1.

———. "Rhine Spill Flooded with Frustration." *Chicago Tribune*, November 16, 1986, Section C at 1.

———. "A Sad Basel Offers Dirge for 'Fluvius Rhinus'." *New York Times*, November 16, 1986, Section 1, 18.

———. "Spill's Effects on Rhine May Be Less than Feared." *New York Times*, January 11, 1987, Section 1, 4.

———. "Swiss Probe Rhine Spill Arson Claim." *Chicago Tribune*, November 13, 1986, Section C, 3.

Pearce, Fred. "Greenprint for Rescuing the Rhine." *New Scientist*, June 26, 1993, 25–29.

Pilarski, Laura, and Roon Lewald. "Chemical Spill Ravages the Rhine." *Engineering News-Record*, November 20, 1986, 12.

Renner, Rebecca. "Controversy Clouds Atrazine Studies." *Environmental Science & Technology* 38, no. 6 (March 15, 2004): 107A-108A.

"Rhine Cleanup Follows Meandering Path." *Chemical Week*, February 13, 1980, 32.

"Rhine Pollution Negligence Alleged." *Financial Times*, November 11, 1986, 44.

"Sandoz: The price of pollution." *The Economist* 301 (November 15, 1986): 80–81.

Shelley, Mary. *Frankenstein, or the Modern Prometheus*. New York: Quality Paperback Book Club, 1994.

Schwabach, Aaron. "The Sandoz Spill: The Failure of International Law to Protect the Rhine from Pollution." *Ecology Law Quarterly* 16 (1989): 443, 451.

Souder, William. "It's Not Easy Being Green: Are Weed-Killers Turning Frogs into Hermaphrodites?" *Harper's Magazine* 313 (August 2006), 59–66.

Stigliani, William M., Peter R. Jaffe, and Stefan Anderberg. "Heavy Metal Pollution in the Rhine Basin." *Environmental Science and Technology* 27 (1993): 786–792.

"Swiss Protest Spill." *Washington Post*, November 10, 1986.

"Swiss Wastes to Be 'Perfumed' in France." *Chemical Week*, July 12, 1978, 39.

Templeman, John, Frederic A. Miller, Jonathan Kapstein, and Laura Pilarski. "Suddenly, A Deathwatch on the Rhine," *Business Week*, November 24, 1986, 80–81

"Thirty Tons of Poison Assault River Life." *New York Times*, November 16, 1986, Section 4, 3.

Tomforde, Anna. "Rhine Pollution Now a Disaster." *The Guardian*, November 8, 1986.

———. "Rhinelanders Despair of 'Poisonous Broth.'" *The Guardian*, November 15, 1986.

Tuohy, William. "Four Nations Try to Cope as Rhine Spreads Spilled Chemicals," *Los Angeles Times*, November 13, 1986, Part 1, 1.

"Water Pollution Alert Lifted in Rhine Chemical Poison Spill." *Los Angeles Times*, November 16, 1986, Part 1, 18.

Watson, Russell, Debbie Seward, Ruth Marshall, Scott Sullivan, and Friso Endt. "The Blotch on the Rhine." *Newsweek*, November 24, 1986, 58–60.

"We Thought We Were Better," *The Economist* 301 (November 29, 1986): 43.

Weber, Urs. "The 'Miracle' of the Rhine." *The Courier (UNESCO)*, June 2000.

Wicks, John. "Incomprehensible That Danger Was Not Realised," *Financial Times*, November 15, 1986, 8.

———. "Dominant Force in Local Economy." *Financial Times*, April 5, 1984, Section II, 33.

World Insurance Report, October 14, 1987, and October 30, 1987.

PRINCE WILLIAM SOUND

Alaska Department of Fish and Game. Division of Subsistence and the Chugach Regional Resources Commission. *Subsistence Service Update: Overview of Study Findings of Exxon Valdez Oil Spill Restoration Project* (No. 99471).

Ballachey, Brenda E., James L. Bodkin, and Anthony R. DeGange. "An Overview of Sea Otter Studies." In Loughlin, *Marine Mammals and the Exxon Valdez*, 47–59.

Barringer, Felicity. "$92 Million More Is Sought for Exxon Valdez Cleanup." *New York Times*, June 2, 2006, www.nytimes.com.

Benyshek, Denita. "Journey into the Dead Zone: Interview with Tom Dragt." In Frost, Helen, ed. *Season of Dead Water.* Portland, OR: Breitenbush, 1990, 27–28.

Bodkin, J. L., and B. E. Ballachey. "Sea Otter." In *Restoration Notebook. Exxon Valdez* Oil Spill Trustee Council, November 1997. http://www.oilspill.state.ak.us/otter1.htm. Web site is no longer active—study available from author.

Bonvillain, Nancy. *The Inuit.* Indians of North America, edited by Frank W. Porter III. New York: Chelsea House, 1995.

Cook, Lynn J. "Exxon Mobil Posts Record 4th Quarter Profit of $10.71 Billion." *Houston Chronicle*, January 30, 2006.

Davis, Nancy Y. "The *Exxon Valdez* Oil Spill, Alaska." In *The Long Road to Recovery: Community Responses to Industrial Disaster*, edited by James K. Mitchell, 231–272. New York: United Nations University Press, 1996.

Exxon Valdez Oil Spill Trust Council Web site. http://www.evostc.state.ak.us.

Exxon Valdez Oil Spill Trustee Council. *Then and Now: A Message of Hope—15th Anniversary of the* Exxon Valdez *Oil Spill* (2004).

Fall, James A. "Overview of Research by the Division of Subsistence, Alaska Department of Fish and Game, on the Sociocultural Consequences of the *Exxon Valdez* Oil Spill." In *Proceedings of Social Indicators Monitoring Study Peer Review Workshop*, June 18–19, 1996, Anchorage, Alaska. Prepared for the United States Department of the Interior, Minerals Management Service, Alaska OCS Region. OCS Study MMS 96-0053 (September 1996), 35–54.

———. "Subsistence." In *Restoration Note Book. Exxon Valdez* Oil Spill Trustee Council (September 1999).

Fall, James A., and L. Jay Field. "Subsistence Uses of Fish and Wildlife before and after the *Exxon Valdez* Oil Spill." *American Fisheries Society Symposium* 18 (1996): 819, 820–823.

Fall, James A., Lee Stratton, Philippa Coiley, Louis Brown, Charles J. Utermohle, and Gretchen Jennings. *Subsistence Harvests and Uses in Chenega Bay and Tatitlek in the Year Following the* Exxon Valdez *Oil Spill.* Division of Subsistence, Alaska Department of Fish and Game. Technical Paper No. 199 (August 1996).

Frost, Helen, ed. *Season of Dead Water.* Portland, OR: Breitenbush, 1990.

Frost, Kathryn J. "Harbor Seal: Phoca vitulina richardsi." In *Restoration Notebook.* Exxon Valdez Oil Spill Trustee Council (October 1996).

Greenhouse, Linda. "Justices to Hear Exxon's Challenge to Punitive Damages." *New York Times,* October 30, 2007, www.nytimes.com.

Guterman, Lila. "Slippery Science." *Chronicle of Higher Education* 51 (September 24, 2004): A12-A16.

Holloway, Marguerite. "Sounding Out Science." *Scientific American,* October 1996, 106–112.

In re *Exxon Valdez.* 104 F.3d 1196 (9th Cir. 1997); 270 F.3d 1215, 1244 (9th Cir. 2001).

Keeble, John. *Out of the Channel: The Exxon Valdez Oil Spill in Prince William Sound.* New York: HarperCollins, 1991.

Lebedoff, David. *Cleaning Up: The Story behind the Biggest Legal Bonanza of Our Time.* New York: Free Press, 1997.

Loughlin, Thomas R., ed. *Marine Mammals and the Exxon Valdez.* San Diego: Academic Press, 1994.

Liptak, Adam. "4.5 Billion Award Set For Spill of *Exxon Valdez.*" *New York Times,* January 29, 2004, A18.

———. "Damages Cut Against Exxon in Valdez Case. *New York Times,* June 26, 2008.

Matkin, Craig, and Eva Saulitis. "Killer Whales." In *Restoration Notebook.* Exxon Valdez Oil Spill Trustee Council (1997).

Meganack, Walter, Sr. "When the Water Died." In Frost, *Season of Dead Water,* 6–8.

Morris, Byron F., and Thomas R. Loughlin, "Overview of the *Exxon Valdez* Oil Spill, 1989–1992." In Loughlin, *Marine Mammals and the Exxon Valdez,* 1–22.

National Research Council. "Human Ecology." The Great Alaska Earthquake of 1964. Vol. 7. Washington DC: National Academy of Sciences, 1970, 392–99.

O'Driscoll, Mary. "Oil Spills: U.S., Alaska want more cash for Exxon Valdez cleanup." *Energy & Environment News,* June 1, 2006.

Ott, Dr. Riki. "Oil and the Marine Environment." In *Prince William Sound Environmental Reader: 1989 T/V Exxon Valdez Oil Spill,* edited by Nancy R.

Lethcoe and Lisa Nurnberger, 30–35. Valdez, AK: Prince William Sound Conservation Alliance, 1989.

Pain, Stephanie. "Species after Species Suffers from Alaska's Spill." *New Scientist*, February 13, 1993, 5.

Palinkas, Lawrence A., Michael A. Downs, John S. Petterson, and John Russell, "Social, Cultural and Psychological Impacts of the *Exxon Valdez* Oil Spill." *Human Organization* 52 (Spring 1993): 1–13.

Parrish, Julia K., and P. Dee Boersma. "Muddy Waters." *American Scientist* 83 (March-April 1995), 112–115.

Peterson, Charles H., Stanley D. Rice, Jeffrey W. Short, Daniel Esler, James L. Bodkin, Brenda E. Ballachey, and David B. Irons. "Long-Term Ecosystem Response to the Exxon Valdez Oil Spill." *Science*, December 19, 2003, 2082–2086.

Porretto, John. "Exxon Mobil Reports Record $45.2 Billion Profit for 2008." *New York Times*, Jnuary 30, 2009.

Raloff, Janet. "Brain lesion helps explain seal loss." *Science News*, February 20, 1993, 126.

———. "A (Killer) whale of a mystery." *Science News*, February 20, 1993, 126.

———. "An Otter Tragedy: Understanding the Sea Otter's Vulnerability to Oil Has Proved Costly to All Involved." *Science News*, March 27, 1993, 200–202.

Seitz, Jody, and Rita Miraglia. "Chenega Bay." In *An Investigation of the Sociocultural Consequences of Outer Continental Shelf Development in Alaska*, edited by James A. Fall and Charles J. Utermohle. Technical Report No. 160. Submitted by the Division of Subsistence of the Alaska Department of Fish and Game to the United States Department of the Interior (March 1995).

Senkowsky, Sonya. "The Oil and the Otter." *Scientific American*, May 2004, 30–32.

Short, Jeffrey W., Mandy R. Lindeberg, Patricia M. Harris, Jacek M. Maselko, Jerome J. Pella, and Stanley D. Rice. "Estimate of Oil Persisting on the Beaches of Prince William Sound 12 Years after the *Exxon Valdez* Oil spill." *Environmental Science & Technology* 38 (January 1, 2004): 19–25.

Smith, Conrad. *Media and Apocalypse: News Coverage of the Yellowstone Forest Fires, Exxon Valdez Oil Spill, and Loma Prieta Earthquake.* Westport, CT: Greenwood, 1992.

St. Aubin, David J., and Joseph R. Geraci. "Summary and Conclusions." In Loughlin, *Marine Mammals and the* Exxon Valdez, 374.

Wells, P. G., J. N. Butler and J. S. Hughes. "Introduction, Overview, Issues," 3–23. Exxon Valdez *Oil Spill: Fate and Effects in Alaskan Waters.* Philadelphia: American Society for Testing and Materials, 1995.

Wheelwright, Jeff. *Degrees of Disaster Prince William Sound: How Nature Reels and Rebounds.* New York: Simon & Schuster, 1994.

Wiens, John A. "Oil, Seabirds, and Science," *Bioscience* 46:586–97.

Wooley, Christopher B. "Alutiiq Culture before and after the *Exxon Valdez* Oil Spill." *American Indian Culture and Research Journal* 19 (1995): 125–53.

KUWAIT

Abbott, Alison. "WHO Plans Study of Gulf War Fallout." *Nature*, September 13, 2001, 97.

Abdali, Fatima, and Sami Al-Yakoob. "Environmental Dimensions of the Gulf War: Potential Health Impacts." In *The Gulf War and the Environment*, edited by Farouk El-Baz and R. M. Makharita, 85–113. Lausanne, Switzerland: Gordon and Beach Science, 1994.

Abuzinda, Abdulaziz and Friedhelm Krupp. "What Happened to the Gulf: Two Years after the World's Greatest Oil-Slick." *Arabian Wildlife* 2 (1994). www.arabianwildlife.com/archive/vol2.1/oilglf.htm.

Al-Hassan, Jassim. "A Personal Perspective." In Al-Shatti and Harrington, *The Environmental and Health Impact of the Kuwait Oil Fires*, 65.

Al-Hassan, Jassim M. "The Iraqi Invasion—Environmental Catastrophe." In Al-Shatti, Ahmed K.S. and J. M. Harrington, eds. *The Environmental and Health Impact of the Kuwait Oil Fires*. Proceedings of an International Symposium, October 17, 1991. Birmingham, U.K.: The University of Birmingham, 1992, 6.

Al-Shatti, Ahmed K. S. and J. M. Harrington, eds. *The Environmental and Health Impact of the Kuwait Oil Fires*. Proceedings of an International Symposium, October 17, 1991. Birmingham, U.K.: The University of Birmingham, 1992.

Arbuthnot, Felicity. "Deserted Victims of War." *The Ecologist* 30 (September 2000): 58–59.

Avril, Tom. "The Impact on the Environment Is Often Devastating." *Philadelphia Inquirer*, March 3, 2003.

Bakan, S., A. Chlond, U. Cubasch, J. Feichter, H. Graf, H. Grassl, K. Hasselmann, I. Kirchner, M. Latif, E. Roeckner, R. Sausen, U. Schlese, D. Schriever, I. Schult, U. Schumann, F. Sielmann, and W. Wellke. "Climate Response to Smoke from the Burning Oil Wells in Kuwait." *Nature*, May 30, 1991, 367–371.

Bloom, Saul, John M. Miller, James Warner, and Philippa Winkler, eds. *Hidden Casualties: Environmental, Health and Political Consequences of the Persian Gulf War*. Berkeley, CA: North Atlantic, 1994.

British Ministry of Defense website on the Gulf War. http://mod.uk/issues/gulfwar.

Brown, Phil, Stephen Zavestoski, Sabrina McCormick, Meadow Linder, et al. "A Gulf of Difference: Disputes over Gulf War-Related Illnesses." *Journal of Health and Social Behavior* 42 (September 2001): 235–257.

Ciment, James. "Iraq Blames Gulf War Bombing for Increase in Child Cancers." *British Medical Journal* 317 (December 12, 1998): 1612.

Collier, Robert. "Iraq Links Cancers to Uranium Weapons." *San Francisco Chronicle*, January 13, 2003.

Daehler, Curtis C., and Shyamal K. Majumdar. "Environmental Impacts of the Persian Gulf War." In *Natural and Technological Disasters: Causes, Effects, and Preventive Measures*, edited by S. K. Majumdar, G. S. Forbes, E. W. Miller, and R. F. Schmalz. Easton, PA: Pennsylvania Academy of Science, 1992.

Daoud, Dr. M. W. "Ahmadi City under the Smoke." In Al-Shatti and Harrington, *The Environmental and Health Impact of the Kuwait Oil Fires*, 27.

El Desouky, Dr. Mustafa, and Dr. Mahmood Y. Abdulraheem. "Impact of Oil Well Fires on the Air Quality in Kuwait." In Al-Shatti and Harrington, 16–26.

Great Barrier Reef Marine Park Authority of Australia. http://www.gbrmpa.gov.au.

Green Cross International. *An Environmental Assessment of Kuwait Seven Years After the Gulf War* (1998): III-V, 13–20, 46–58, 71–74.

Hahn, Jürgen. "Environmental Effects of the Kuwaiti Oil Field Fires." *Environmental Science Technology* 25 (1991): 1531–1532.

Hawley, T. M. *Against the Fires of Hell: The Environmental Disaster of the Gulf War.* New York: Harcourt Brace Jovanovich, 1992.

Hobbs, Peter V. "Introduction." *Journal of Geophysical Research* 97 (September 20, 1992): 14, 481.

Hobbs, Peter V., and Lawrence F. Radke, "Airborne Studies of the Smoke from the Kuwait Oil Fires." *Science*, May 15, 1992, 987–991.

Horgan, John. "U.S. Gags Discussion of War's Environmental Effects." *Scientific American*, May 1991, 24.

Husain, Tahir. *Kuwaiti Oil Fires: Regional Environmental Perspectives.* Tarrytown, NY: Elsevier Science, 1995.

"Interview with Burr Heneman," In Bloom et al., *Hidden Casualtiesr*, 53–62.

"Interview with Mike Evans." In Bloom et al., *Hidden Casualties*, 63–68.

"Iraq: Scientists to study risks from DU shells," *Greenwire*, April 22, 2003.

Kapp, Clare. "WHO Sends Team to Iraq to Investigate Effects of Depleted Uranium." *The Lancet*, September 1, 2001, 737.

Kemp, Penny. "For Generations to Come: The Environmental Catastrophe." In *Beyond the Storm: A Gulf Crisis Reader*, edited by Phyllis Bennis and Michel Moushabeck, 325–334. New York: Olive Branch, 1991.

Klare, Michael T. *Resource Wars: The New Landscape of Global Conflict.* New York: Henry Holt, 2001.

Limaye, S. S., V. E. Suomi, C. Velden, and G. Tripoli. "Satellite Observations of Smoke from Oil Fires in Kuwait." *Science*, June 14, 1991, 1536–1539.

Mesler, Bill. "The Pentagon's Radioactive Bullet." *The Nation*, October 12, 1996, 11.

Miller, John M.. "Chronology of a Coverup." In Bloom et al., *Hidden Casualties*, 92–94.

Montague, Peter. "Depleted Uranium Weapons of War." *Rachel's Democracy & Health News* 788 (April 1, 2004). http://www.rachel.org.

North, Andrew. "Leukemia in Iraq." *Washington Report on Middle East Affairs* 19 (July 2000): 29.

Peterson, Scott. "DU's Fallout in Iraq and Kuwait: A Rise in Illness?" *Christian Science Monitor*, April 29, 1999, 14.

Pilger, John. "Iraq: The Great Cover-up." *New Statesman*, January 22, 2001, 8.

Pope, C. Arden, III, Richard T. Burnett, Michael J. Thun, Eugenia E. Calle, Daniel Krewski, Kazuhiko Ito, George D. Thurston. "Lung Cancer, Cardiopulmonary Mortality, and Long-Term Exposure to Fine Particulate Air Pollution." *Journal of the American Medical Association* 287 (March 6, 2002): 1132–1141.

Preen, Anthony, Helene Marsh, and George E. Heinsohn. "Dugongs in the Red Sea and Persian Gulf." *Yemen Update* 33 (1993): 27.

Raloff, J., and R. Monastersky. "Gulf Oil Threatens Ecology, Maybe Climate." *Science News*, February 2, 1991, 71–73.

Research Advisory Committee on Gulf War Veterans' Illnesses. *Gulf War Illness and the Health of Gulf War Veterans: Scientific Findings and Recommendations.* Washington, DC: U.S Gov't Printing Office, 2008.

Reynolds, John E., III, and Daniel K. Odell. *Manatees and Dugongs.* New York: Facts on File, 1991.

Schneider, Stephen H. "Smoke Alarm." *World Monitor* 4 (March 1991): 50–51.

Sharratt, M., and Margaret Butler. "Toxicological Effect of Oil Smoke." In Al-Shatti and Harrington, *The Environmental and Health Impact of the Kuwait Oil Fires*, 51–56.

Small, Richard D. "Environmental Impact of Fires in Kuwait." *Nature*, March 7, 1991, 11–12.

Shimoyachi, Nao. "Ex-Military Doctor Decries Use of Depleted Uranium Weapons." *Japan Times*, November 22, 2003, 3.

Snashall, David. "Smoke and Health—Assembling the Jigsaw in Kuwait." In Al-Shatti and Harrington, *The Environmental and Health Impact of the Kuwait Oil Fires*, 44.

Steadman, Tom. "Desert Storm: Still Raging; Vets Say They Went Healthy, Came Back Sick." *Greensboro News & Record*, November 11, 2001.

Tager, Jeremy. "Going, Going, Dugong." *Earth Island Journal* 14 (Summer 1999): 36.

Turco, R. P., O. B. Toon, T. P. Ackerman, J. B. Pollack, C. Sagan. "Climate and Smoke: An Appraisal of Nuclear Winter." *Science*, January 1990, 166–175.

Warner, Sir Frederick. "The Environmental Consequences of the Gulf War." *Environment* 33 (June 1991): 7 ff.

Wheelwright, Jeff. *The Irritable Heart: The Medical Mystery of the Gulf War.* New York: Norton, 2001.

Whitaker, Brian. "Society: Environment: The Black Desert: For Kuwait, Vast Lakes of Oil, Contaminated Water Reserves and Increasing Cases of Asthma Are the Legacies of the Gulf War." *The Guardian*, August 16, 2000.

DASSEN AND ROBBEN ISLANDS

Allwright, David, and Heidi Stout. "Losses and Mortalities Following the Treasure Oil Spill." *Proceedings Treasure Oil Spill Conference on Wildlife Issues*, November 23–25, 2000. http://www.capenature.org.za/what_is_new/treasure/allwright.html. Website is no longer active—study available from author.

"Biggest Rescue Ever' Helps Penguins Survive Oil Spill." *Houston Chronicle*, October 12, 2000.

Chester, Jonathan. *The Nature of Penguins*. Berkeley, CA: Celestial Arts, 2001.

Crawford, R. J. M., S. A. Davis, R. Harding, L. F. Jackson, T. M. Leshoro, M. A. Meyer, R. M. Randall, L. G. Underhill, L. Upfold, A. P. Van Dalsen, E. Van Der Merwe, P. A. Whittington, A. J. Williams, and A. C. Wolfaardt. "Initial Effects of the *Treasure* Oil Spill on Seabirds off Western South Africa." Avian Demography Unit, University of Cape Town. http://web.uct.ac.za/depts/stats/adu/oilspill/oilspill.htm.

Crawford, R. J. M., M. A. Meyer, L. G. Underhill, and L. Upfold. "Relocation of African Penguins to Prevent Their Becoming Oiled after the Sinking of the *Treasure*, and the Tracking of Their Return." *Proceedings Treasure Oil Spill Conference on Wildlife Issues*, November 23–25, 2000. http://www.capenature.org.za/what_is_new/treasure/crawford.html. Web site is no longer active—study available from author.

Fick, Tim. "Penguin Cleaning: Five Hours in the Life of a First-time Volunteer." http://www.geoscience.org.za/bellville/penguin.html. Website is no longer active—study available from author.

Ford, Mike. "First Impressions of a SANCCOB Volunteer" and "Second Impressions of a SANCCOB Volunteer." Avian Demography Unit, University of Cape Town. http://web.uct.ac.za/depts/stats/adu/oilspill/mikeford.htm and http://web.uct.ac.za/depts/stats/adu/oilspill/mikefor2.htm.

Glazewski, Jan, and Emma Dingle. "*Treasure* Oil Spill—Legal lessons learnt." *Proceedings* Treasure *Oil Spill Conference on Wildlife Issues*, November 23–25, 2000. http://www.capenature.org.za.

Gosling, Melanie. "Island Stake-out Succeeds in Retrieving 'Backpack' from Pamela the Elusive Penguin." Avian Demography Unit, University of Cape Town. http://www.uct.ac.za/depts/stats/adu/oilspill/pamela.htm

Hunt, Steven. "Mopping up with Nature's Help." http://www.exn.ca. Web site is no longer active—study available from author.

"International Effort Required to Combat Oil Spills." Iafrica.com. June 21, 2001. http://iafrica.com/pls/procs/SEARCH.ARCHIVE?p_content_id=416496&p_site_id=2.

International Fund for Animal Welfare (IFAW). *Spill: Saving Africa's Oiled Penguins*. South Africa: Inyati, 2000.

"Jackass Penguin." http://www.botany.uwc.ac.za.

Line, Les. "Into the Abyss." *International Wildlife* 27 (September/October 1997): 12.

Lynch, Wayne. *Penguins of the World*. Buffalo, NY: Firefly, 1997.

Maykuth, Andrew. "Volunteers Flock to Aid Penguins." (Montreal) *Gazette*, July 2, 2000, A11.

"New Hope for Endangered African Penguins, Conservationists Welcome Clean-up Reimbursement, Rebounding Species." *PR Newswire*, November 30, 2001.

"Percy Becomes the First Penguin Pin-up." *Proceedings* Treasure *Oil Spill Conference on Wildlife Issues*, November 23–25, 2000. http://www.capenature.org.za/what_is_new/pinup.html. Web site is no longer active—study available from author.

"Percy the Penguin Passes Checkup after Swim Home off South Africa; He Was One of Thousands Rescued from Oil Spill." *St. Louis Post-Dispatch*, July 30, 2000.

Reilly, Pauline. *Penguins of the World*. New York: Oxford University Press, 1994.

Robben Island. www.freedom.co.za.

"Satellite Tag Tracks Peter the Penguin on His Swim Home." *Geographical Magazine* 72 (September 2000): 10.

Scarth, Sarah. "South African Oiled Penguin Rescue Scores Record Results." *PR Newswire*, August 16, 2000.

Singer, Rena. "South Africa's Oil-soaked Penguins Get a Scrubbing." *Christian Science Monitor* 92 (July 26, 2000): 7.

Skelton, Renee. "The Great Penguin Rescue." *National Geographic World* 305 (January 2001): 12.

"Treasure Spill Report." International Bird Rescue Research Center.

Underhill, Les. "A Brief History of Penguin Oiling in South African Waters." Avian Demography Unit, University of Cape Town. http://www.uct.ac.za/depts/stats/adu/oilspill/oilhist.htm.

———. "The *Treasure* Oil Spill." Avian Demography Unit, University of Cape Town. http://www.uct.ac.za/depts/stats/adu/oilspill/diary.htm.

Visagie, Johan. "Treasure Oil Spill—1 Year Later: Census Proves That Penguins Are Making a Remarkable Recovery." In *Proceedings Treasure Oil Spill Conference on Wildlife Issues*, November 23–25, 2000. http://www.capenature.org.za. Web site is no longer active—study available from author.

Wines, Michael. "Robben Island Journal: Dinner Disappears, and African Penguins Pay the Price." *New York Times*, June 4, 2007, www.nytimes.com.

Wolfaardt, Anton. "The Capture and Removal of Clean Penguins from Dassen Island." In *Proceedings* Treasure *Oil Spill Conference on Wildlife Issues*,

November 23–25, 2000. http://www.capenature.org.za/what_is_new/treasure/capture.html. Web site is no longer active—study available from author.

Wolfaardt, Anton. "Information about the African Penguin." Western Cape Nature Conservation Board, Dassen Island Nature Reserve. http://www.cmc.gov.za/pht/Treasure/AboutPenquins.htm. Web site is no longer active—study available from author.

RAINFOREST

"Amazon Destruction: Why Is the Rainforest Being Destroyed in Brazil?" http://rainforests.mongabay.com/amazon/amazon_destruction.html (March 4, 2008).

"Asbury Park, NJ, Faces Protest in Plan to Redo Boardwalk with Tropical Wood." (Newark) *Star-Ledger*, January 16, 2001.

Assies, Willem. *Going Nuts for the Rainforest: Non-Timber Forest Products, Forest Conservation and Sustainability in Amazonia*. Amsterdam: Thela, 1997.

Bierregaard, Richard O. Jr., Claude Gascon, Thomas E. Lovejoy, and Rita C. G. Mesquita. *Lessons from Amazonia*. New Haven, CT: Yale University Press, 2001.

Bunyard, Peter. "A Stake through the Heart of the World." *The Ecologist*, July/August 2005, 30–35.

Carneiro, Robert L. "Indians of the Amazonian Forest." In Denton, Julie Sloan, and Christine Padoch, eds. *People of the Tropical Rain Forest*. Berkeley and Los Angeles: University of California Press, 1988, 73–86.

Caufield, Catherine. *In the Rainforest: Report from a Strange, Beautiful, Imperiled World*. Chicago: University of Chicago Press, 1991.

Clay, Jason W. "Indigenous Peoples: The Miner's Canary for the Twentieth Century." In Head, Suzanne, and Robert Heinzman, eds. *Lessons of the Rainforest*. San Francisco: Sierra Club, 1990, 106–117.

Conrad, Joseph. *The Heart of Darkness*. In *Three Tales by Joseph Conrad*, with a general introduction by Albert J. Guerard. New York: Dell, 1960.

DeCurtis, A., and H. Ritts. "Sting." *Rolling Stone*, February 7, 1991, 42.

Denton, Julie Sloan, and Christine Padoch, eds. *People of the Tropical Rain Forest*. Berkeley and Los Angeles: University of California Press, 1988.

———. "The Tropical Rain-Forest Setting." In Denton and Padoch, *People of the Tropical Rain Forest*, 29, 32–34.

Dwyer, Augusta. *Into the Amazon: The Struggle for the Rain Forest*. San Francisco: Sierra Club, 1990.

Eden, Michael J. *Ecology and Land Management in Amazonia*. New York: Belhaven, 1990.

"For US Company, Tribe Partnership Is Bottom Line." *Christian Science Monitor* 89 (November 11, 1997): 8.

Foroohar, Rana. "The New Green Game." *Newsweek*, August 27, 2001, 62.

Forest Stewardship Council. United States. http:/www.fscus.org.

Forsyth, Adrian. "Salts of the Earth: Nutrient Cycles in Tropical Forests," 68–73. In *Portraits of the Rainforest.* Ontario: Camden House, 1990.

Head, Suzanne, and Robert Heinzman, eds. *Lessons of the Rainforest.* San Francisco: Sierra Club, 1990.

Hecht, Susanna, and Alexander Cockburn. *The Fate of the Forests: Developers, Destroyers and Defenders of the Amazon.* New York: HarperCollins, 1990.

Hennigan, Tom. "Rancher Jailed for Ordering Murder of Rainforest Nun." *New York Times,* May 17, 2007, www.nytimes.com.

Lamb, Christina. "Sting's Amazon Tribe in Peril as Miners Return." *New York Times,* October 14, 2007, www.nytimes.com.

Laurance, William F., Mark A. Cochrane, Scott Bergen, Philip M. Fearnside, Patricia Delamonica, Christopher Barber, Sammya D'Angelo and Tito Fernandes. "The Future of the Brazilian Amazon." *Science Magazine,* February 1, 2001, 438.

Lewis, Scott, with the Natural Resources Defense Council. *The Rainforest Book: How You Can Save the World's Rainforests.* Los Angeles: Living Planet, 1990.

McCuen, Gary E. *Ecocide and Genocide in the Vanishing Forest: The Rainforests and Native People.* Hudson, WI: GEM, 1993.

Meggers, Betty J. "The Prehistory of Amazonia." In Denton and Padoch, *People of the Tropical Rain Forest,* 54.

Moore Lappé, Frances, and Rachel Schnerman. "Taking Population Seriously: Power and Fertility." In Head and Heinzman, *Lessons of the Rainforest,* 131–144.

Morton, David "Looking at Lula." *E Magazine,* September/October 2005, 14–16.

Perry, Donald R. "Tropical Biology: A Science on the Sidelines." In Head and Heinzman, *Lessons of the Rainforest,* 25–36.

Putz, Francis E., and N. Michelle Holbrook. "Tropical Rain-Forest Images." In Denton and Padoch, *People of the Tropical Rain Forest,* 45.

"Rainforest Facts." www.rain-tree.com/facts.htm, March 4, 2008.

Reiss, Bob. *The Road to Extrema.* New York: Summit, 1992.

Revkin, Andrew. *The Burning Season: The Murder of Chico Mendes and the Fight for the Amazon Rain Forest.* Boston: Houghton Mifflin, 1990.

———. "Remembering Chico Mendes." *E Magazine,* March/April 2005, 23–25.

Rich, Bruce. "Multilateral Development Banks and Tropical Deforestation." In Head and Heinzman, *Lessons of the Rainforest,* 118–130.

Rohter, Larry. "Relentless Foe of the Amazon Jungle: Soybeans." *New York Times,* September 17, 2003, A3.

———. "Deep in the Amazon, Vast Questions about the Climate." *New York Times,* November 4, 2003, F1.

———. "Loggers, Scorning the Law, Ravage the Amazon Jungle." *New York Times,* October 16, 2005, 6.

"Satellite Surveillance Curbs Illegal Logging—Brazilian minister." *Greenwire,* November 12, 2003.

Schmink, Marianne. "Big Business in the Amazon." In Denton and Padoch, *People of the Tropical Rain Forest*, 168–172.

Smith, Tony. "Rain Forest Is Losing Ground Faster in Amazon, Photos Show." *New York Times*, June 28, 2003, A2.

"Dorothy Stang." *Wikipedia*. http://en.wikipedia.org/wiki/Dorothy_Stang.

Taylor, Kenneth Iain. "Why Supernatural Eels Matter." In Head and Heinzman, *Lessons of the Rainforest*, 184–195.

Weart, Spencer R. *The Discovery of Global Warming.* Cambridge, MA: Harvard University Press, 2003.

Several websites provide current statistics on the tropical rain forests, including the Rainforest Alliance: http:/www.rainforest-alliance.org; and the Rainforest Action Network: http:/www.ran.org.

GLOBAL CLIMATE CHANGE

Ahlstrom, Dick. "Ireland Shrinks by 750 Acres a Year as Sea Eats Shoreline." *Irish Times*, March 26, 2002, 3.

Ahlstrom, Dick, and Frank McNally. "Urgent Need to Prepare for Climate Change." *Irish Times*, February 20, 2001, 9.

Allen, Leslie. "Will Tuvalu Disappear beneath the Sea? Global Warming Threatens to Swamp a Small Island Nation." *Smithsonian* 35 (August 2004): 44–52.

"Atlantic Ocean Becoming Saltier, Evaporating Faster—Study." *Greenwire*, December 18, 2003.

Barringer, Felicity. "California, Taking Big Gamble, Tries to Curb Greenhouse Gases." *New York Times*, September 15, 2006, A1, A20–21.

Bloomfield, Janine, Molly Smith and Nicholas Thompson. *Hot Nights in the City: Global Warming, Sea-Level Rise and the New York Metropolitan Region.* Environmental Defense Fund, 1999.

Carey, John, and Sarah R. Shapiro. "Global Warming: Why Business Is Taking It So Seriously." *Business Week*, August 16, 2004, 60–69.

Claussen, Eileen. "Climate Change: The Political Challenges." *Pew Center on Global Climate Change*. www.pewclimate.org.

———. "Emission Reductions: Main Street to Wall Street." *Pew Center on Global Climate Change*. July 17, 2002. http://www.pewclimate.org/press_room.

Climate Change and a Global City: The Potential Consequences of Climate Variability and Change—Metro East Coast. A Report of the Columbia Earth Institute, for the U.S. Global Change Research Program, July 2001.

Dunn, Seth. *Reading the Weathervane: Climate Policy From Rio To Johannesburg.* Worldwatch Paper 160. Washington, DC: The Worldwatch Institute, August 2002.

Egan, Timothy. "Warmth Transforms Alaska, and Even Permafrost Isn't." *New York Times*, June 16, 2002, A1, A18.

Epstein, Paul R. "Global Chilling." *New York Times*, Op-Ed, January 28, 2004, A25.

Forero, Juan, "As Andean Glaciers Shrink, Water Worries Grow." *New York Times*, November 24, 2002, 3.

Freeman, Andrew. "Danger threshold for global temperature rise is real, scientists say." *Greenwire*, December 12, 2003.

———. "IPCC projections may have underestimated rate, effects of warming, scientists say." *Greenwire*, October 25, 2004.

Frey, Darcy. "How Green Is BP?" *New York Times Magazine*, December 8, 2002, 99.

Geman, Ben. "Saying Science is 'Clear,' Shell's Chief Urges Action on Emissions." *Greenwire*, February 8, 2006.

Glick, Daniel. "The Big Thaw." *National Geographic* 206 (September 2004): 13–33.

Goodstein, Laurie. "Evangelical Leaders Join Global Warming Initiative: Legislation Is Urged to Reduce Emissions." *New York Times*, February 8, 2006, A12.

Gore, Al. *An Inconvenient Truth: The Planetary Emergency of Global Warming and What We Can Do About It*. Emmaus, PA: Rodale Press, 2006.

Gray, John. "The Global Delusion." *New York Review of Books*, April 27, 2006, 20–23.

Greenwire, an environmental e-newletter, August 7, 2003.

Hansen, James, Makiko Sato, Pushker Kharecha, David Beerling, Valerie Masson-Delmotte, Mark Pagani, Maureen Raymo, Dana L. Royer, and James C. Zachos. "Target Atmospheric CO2: Where Should Humanity Aim?" Latest draft posted on Hansen website, April 7, 2008. www.columbia.edu/~jehl/

Hansen, Jim. "The Threat to the Planet." *New York Review of Books*, July 13, 2006, 12–16.

Houghton, John. *Global Warming: The Complete Briefing*. 2nd ed. Cambridge: Cambridge University Press, 1997.

Inhofe, Senator James. Chairman of the Committee on Environment and Public Works, July 28, 2003. Speech on the Senate floor. Web site for U.S. Senate Committee on Environment and Public Works. http://epw.senate.gov.

Intergovernmental Panel on Climate Change (IPCC) Reports can be found at /www.ipcc.ch/

"IPCC Meeting in Paris to Prepare for Next Climate Change Report." Web site for the Intergovernmental Panel on Climate Change, February 14, 2003. http://www.ipcc.ch/press.

Kaplun, Alex. "Experts argue environmental protection is good for business." *Greenwire*, December 4, 2003.

Kennedy, Robert F., Jr. "Better Gas Mileage, Greater Security." Frugalmarketing.com, http://www.frugalmarketing.com/dtb/kennedy.shtml.

Kolbert, Elizabeth. *Field Notes from a Catastrophe*. New York: Bloomsbury Publishing, 2006.

Leggett, Jeremy. *The Carbon War.* New York: Routledge, 2001.

"Lucky Break: We Can't Rely on Accidental Discoveries for Vital Information about the Planet." Editorial. *New Scientist,* September 3, 1997, 3.

Martin, Douglas. "Report Warns New York of Perils of Global Warming." *New York Times,* June 30, 1999, 7.

McKibben, Bill. "Changing the Climate-Change Climate." *Grist Magazine,* January 25, 2005. www.grist.org/comments/dispatches/2005/01/25/mckibben

———. "The Coming Meltdown." *New York Review of Books* 53 (January 12, 2006). www.nybooks.com/articles/18616

McKibben, Bill. *The End of Nature.* 2nd ed. New York: Doubleday, 1999.

———. "No More Mr. Nice Guy—Climate Change Is Pushing This Easygoing Enviro over the Edge." *Grist Magazine,* January 12, 2006. www.grist.org/comments/soapbox/2006/01/12/mckibben

———. "Warning on Warming." *New York Review of Books* 54 (March 15, 2007), 44–45

Morello, Lauren. "Polls Find Groundswell of Belief in, Concern about Global Warming." *Greenwire,* April 21, 2006.

Northrop, Michael. "Early Reducers." *The Environmental Forum* 21 (March/April 2004): 16–29.

Northrop, Michael, and David Sassoon. "The Mythology of Economic Peril." *Environmental Finance,* June 2005, 18–19.

O'Driscoll, Mary. "Global Warming to Be a 'Centerpiece' of 2008 Campaign, Sen. Carper Predicts." *Greenwire,* February 8, 2006.

Olsen, Jan M. "Retreating Glaciers Worrying Greenlanders." *Boston Globe,* September 11, 2005. http://www.boston.com

Parson, Edward A., Lynne Carter, Patricia Anderson, Bronwen Wang, and Gunter Weller. "Potential Consequences of Climate Variability and Change for Alaska." In *Climate Change Impacts on the United States: The Potential Consequences of Climate Variability and Change,* by the National Assessment Synthesis Team, 283–312. New York: Cambridge University Press, 2001.

Pew Center on Global Climate Change. *Agenda for Climate Action.* February 2006. http://www.pewclimate.org.

Pew Center on Global Climate Change. *Beyond Kyoto: Advancing the international effort against climate change.* December 2003. http://www.pewclimate.org.

Pew Center on Global Climate Change. *Observed Impacts of global climate change in the U.S.* November 2004. http://www.pewclimate.org.

"Rebuked on Global Warming." Editorial, *New York Times,* March 1, 2003, A18.

"Report: Global Warming Is Shrinking Ireland." *Associated Press,* March 25, 2002.

Revkin, Andew C. "Can Global Warming Be Studied Too Much?" *New York Times,* December 3, 2002, F1, F4.

———. *Global Warming: Understanding the Forecast.* New York: Abbeville, 1992.

———. “Into Thin Air: Kyoto Accord May Not Die (Or Matter).” *New York Times,* December 3, 2003, A6.

———. “No Escape: Thaw Gains Momentum.” *New York Times,* October 25, 2005, F1.

———. *The North Pole Was Here: Puzzles and Perils at the Top of the World.* Boston: Kingfisher, A New York Times Book, 2006.

———. “Ozone Layer Is Improving, According to Monitors.” *New York Times,* July 30, 2003, A11.

———. “Study Finds Warming Trend in Arctic Linked to Emissions.” *New York Times,* October 29, 2004. http://www.nytimes.com.

“Rising Temps Could Kill 25 Percent of World’s Species, Scientists Say.” *Greenwire,* January 8, 2004.

Samuelsohn, Darren. “Northeastern States Sign Regional Global Warming Pact.” *Greenwire,* December 20, 2005.

Stempeck, Brian. “No Easy Road to Reduce Gasoline Consumption—CBO Study.” *Greenwire,* January 7, 2004.

Stern, Nicholas. *The Stern Review on the Economics of Climate Change.* www.hm-treasury.gov.uk/independent_reviews/stern_review_economics_climate_change/sternreview_index.cfm.

Stevens, William K. “New Evidence Finds This Is Warmest Century in 600 Years.” *New York Times,* April 28, 1998, F3.

Thomas, Otti. “Global Warming Poses Flood Threat to Dutch.” *Reuters,* December 1, 1999.

“U.K. Plans Ambitious Program to Meet Emissions Reduction Targets.” *Greenwire,* November 15, 2005.

U.S. Department of Energy and U.S. Department of Environmental Protection Agency. *Fuel Economy Guide: Model Year 2004.* http://www.fueleconomy.gov.

Verrrengia, Joseph B. “As Sea Levels Rise, Way of Life Retreats.” *Los Angeles Times,* September 15, 2002.

Walker, Gabrielle. “Fresh Blow for Greenhouse Skeptics.” *New Scientist* 146 (April 22, 1995): 16.

“Warming Leads to Rise in Fresh Water Entering Arctic Ocean—Report.” *Greenwire,* January 21, 2005.

Weart, Spencer R. *The Discovery of Global Warming.* Cambridge, MA: Harvard University Press, 2003.

Whitty, Julia. “The Thirteenth Tipping Point.” *Mother Jones* 31 (December 2006): 44–51, 100–101.

Worster, Donald. *Dust Bowl: The Southern Plains in the 1930s.* New York: Oxford University Press, 1979.

INDEX

A
abortions, 16, 45, 54, 120. *see also* spontaneous abortions
acetaldehyde, 9, 12–13, 18, 20, 21, 26
Agent Orange, 46, 92, 94, 161
Akazaki, Satoru, 15
Alaska Native Claims Settlement Act (ANCSA), 138–139
Alaska Statehood Act (1958), 138
Allegretti, Mary Helene, 176
Alyeska pipeline, 139–140, 142–143
Apollo Sea, 165–166
arsenic trichloride, 62–63, 67
Atomic Energy Act (1946), 39, 43
atrazine, 132–133

B
Beckett, Margaret, 187
benzene hexachlorophene (BHC), 69
Bertazzi, Pier Alberto, 58
Bhopal, India, 2, 3, 5, 101–108, 157
  aftermath, 104–105
  description of incident at, 101–104
  legal repercussions, 107–108
  methyl isocyanate (MIC) and, 101–103, 105–106
  UCC and, 105–108
Bliss, Russell, 92–95, 97–98, 99–100
Brandt, Willy, 132
Brazilian rainforest, 171–178
  attempts to conserve, 176–178
  destruction of, 173–176
  empate movement, 176–177
  native population, 172–173
  overview, 171–172
British Atomic Energy Authority, 41, 43
British Clean Air Act (1956), 6, 36
British National Radiological Protection Board (NRPB), 118
British Petroleum (BP), 184, 187
Bulka, Peter, 70–71, 73, 78, 79
Bush, George H.W., 156, 183
Bush, George W., 186–187

C
cap-and-trade systems, 187
Cape Nature Conservation, 166
Carbon Club. *see* Global Climate Coalition
Carey, Hugh, 75
Carter, Jimmy, 75–77, 88
Catholic Church, 54
Cavallaro, Aldo, 49
Center for Health, Environment & Justice, 78, 190
Centers for Disease Control (CDC), 93–95, 98
Cerrillo, Debbie, 72, 75–76, 78, 79
Chenega, 139–140, 145, 151–154
Chernobyl, Ukraine, 1–3, 5–6, 89, 109–127, 133, 179–180
  aftermath, 113–115
  Austria and, 118
  description of incident at, 109–110
  evacuations, 119–120
  France and, 118
  government response to incident, 110–113, 120–121
  international reaction to, 117–118
  Lapland and, 117–118

Chernobyl, Ukraine—*Continued*
  lasting effects of, 120–126
  Sweden and, 115–116
  UK and, 118–119
China Syndrome, The, 81–82, 85
Chisso, 9–15, 18, 20–22, 24–29
chloracne, 46, 52–53, 57–58, 67–68, 94
chlorofluorocarbons (CFCs), 182
Chugach Eskimo, 137–139, 146, 180
Ciba-Geigy, 56, 132–133
coal, 1, 31–32, 35–37, 111–112
"coffin allowance", 110
Commoner, Barry, 4
Comprehensive Environmental Response, Compensation and Liability Act. *see* Superfund law

D
Dassen and Robben Islands, 6, 163–169, 180, 189
Degrees of Disaster (Wheelwright), 147
Department of Energy, 156
Department of Environmental Conservation (DEC), 71–73
Department of Health (DOH), 73–75
depleted uranium (DU), 161–162
dioxin, 4, 7, 46, 48–59, 68, 78, 94–99, 132
dodecyl mercaptan, 62
Donahue, Phil, 76
DuPont, 187

E
Earth Summit (1992), 183, 184
Eisenhower, Dwight, 43
El Niño, 185
electric automobiles, 188
empates, 176
Endangered Species Act, 180
Environmental Defense Fund, 97
Environmental Protection Agency (EPA)
  Love Canal and, 72–73, 76–77, 79
  Reagan and, 182
  Times Beach and, 97–99
European Union, 7, 56, 132, 186–187
Exxon Valdez spill, 3, 137–154, 157, 160, 180
  causes of wreck, 140–143
  cleanup, 149–154
  environmental effects of, 143–149
  lawsuits against Exxon, 149–154
Ezuno, 17–18, 19, 29

F
fog, 4–6, 31–37, 157
Food and Drug Administration (FDA), 95
Forest Stewardship Council, 178
Forsmark Nuclear Power Plant, 115

G
Gibbs, Lois, ix, x, 73–79
Givaudan, 46, 48–51, 55, 57
global climate change, 179–188
  effects of, 179–180
  greenhouse gases (GHG) and, 185–188
  IPCC and, 183–187
  Kyoto Protocol and, 184–185
  Montreal Protocol and, 182–183
  overview, 179
  scientific research of, 180–182
Global Climate Coalition, 184
globalization, 188
gold mining, 171, 175
Gorbachev, Mikhail, 120–121
Gorsuch, Anne, 97, 99
Grapes of Wrath, The (Steinbeck), 91
Green Action, 29
greenhouse gases (GHG), 179, 183–188
Greenpeace, 56, 108, 133, 136, 181, 189
Gulf War, 3, 160–162
Gulf War disease, 161

H
halons, 182
Hansen, Jim, 188
harbor seals, 147
Harris, Daniel, 97
Hazelwood, Joseph, 140–142, 151
Hempel, Frank, 93–94
hexachlorophene, 46, 92, 95
Hiroshima, 39, 122, 125
Hoffman-La Roche, 46, 48, 50, 55–57
Hoffman-Taff, 92, 94–95
Hooker Electrochemical Company, 61–68, 71–73, 75–76, 78
Hosokawa, Hajime, 10, 12–14, 18, 20, 26, 29

Hussein, Saddam, 155, 157
hybrid automobiles, 188

I
in utero effects of exposure to toxic elements, 59, 120, 125
Indigenous Peoples' Union, 177
Industrie Chimiche Meda Societa Anonima (ICMESA), 45–50, 54–55, 92
Institute for Amazon Studies, 176
Intergovernmental Panel on Climate Change (IPCC), 183, 185–186
International Atomic Energy Agency (IAEA), 125, 126
International Commission for the Protection of the Rhine (ICPR), 135–136
International Fund for Animal Welfare (IFAW), 166, 168, 189
iodine-131, 42, 44, 112, 116, 118, 119
Irukayama, K., 12
Ishimure, Michiko, 14–20, 22–26, 29

K
Kamimura, Tomoko, 23, 27
Kawamoto, Teruo, 23–27, 28
Kayapó, 175, 177
Khodomchuk, Valery, 109
King Hussein, 156
Kuwabara, Shisei, 11, 17, 18, 19, 29
Kuwait, 155, 155–162, 159–160
Kyoto Protocol, 184–186, 187

L
Lapland, 117–118
Lavelle, Rita, 97–99
Lipari Landfill, 78
London, England, 31–37
  arrival of fog, 31–34
  British Clean Air Act (1956), 36–37
  deaths from fog, 36
  effects of fog, 34–35
Love, William, 61
Love Canal, 2, 4–5, 7, 61–79, 180
  Carter and, 75, 77
  DEC and, 71–72
  DOH study of incidents, 73–75
  dumping by Hooker at site, 62–63
  effects on residents, 68–71, 73–75
  EPA and, 72–73, 76–77, 79
  Hooker's knowledge of hazards, 63–64
  relocation of residents, 75–76
  sale to Niagara Falls school board, 64–67
  TCP and, 67–68

M
Macmillan, Harold, 35, 43
Manhattan Project, 62
McKibben, Bill, 188
Mendes, Chico, 176–177
mercury, 2–4, 9, 11–14, 18, 20–22, 26–29, 108, 132, 134–136, 175
methyl isocyanate (MIC), 101–103, 105–106
Metropolitan Edison (Met Ed) Company, 81, 84–85, 88–89
Minamata (Smith and Smith), 28
Minamata, Japan, 4–5, 7, 9–29, 132, 175
  Chisso's denial of wrongdoing, 12–14
  fight for victims' rights, 18–21
  lawsuit against Chisso, 20–29
  pollution of, 9–10
  Pollution Victims Relief Law, 21
  spread of Minamata Disease in, 10–13
Minamata Disease, 12–14, 20–21, 23–24, 28–29, 179
mirex, 71
Moku, 18, 19, 29
Montreal Protocol, 182
M.V. *Treasure*, 165, 169

N
Nagasaki, 39, 122
National Academy of Sciences (NAS), 182
National Defense Research Institute (Sweden), 115
National Oceanographic and Atmospheric Administration (NOAA), 153–154
National Transportation Safety Board (NTSB), 142
Natural Resources Defense Council (NRDC), 176, 190
Nature Conservancy, 177, 190
Niagara Gazette, 71, 73
Niigata City, 20, 22, 24, 26
Northeastern Pharmaceutical and Chemical Company, Inc. (NEPACCO), 92–93, 95–96

Nuclear Regulatory Commission (NRC), 85, 86, 87–88

O
Obama, Barack, 187
open-hearth fires, 32
Orca killer whales, 147–148

P
Paoletti, Paolo, 47, 50, 55
Paradise in the Sea of Sorrow: Our Minamata Disease, 14
Paringaux, Bernard, 56
penguins, 6, 163–169, 180
Penney Report, 43
Phillips, Patrick, 93–94, 95
Piatt, Judy, 93–94, 97–98, 100
Pikalov, Vladimir, 114
Prima Linea, 55
Prince William Sound, Alaska. *see Exxon Valdez* spill
Pripyat, 109–110, 113–115, 122

Q
Queen's University, Belfast, 124

R
rainforests. *see* Brazilian rainforest
Reagan, Ronald, 96–98, 182
Regional Greenhouse Gas Initiative, 187
Reid, Bob, 84
Resource Conservation and Recovery Act (RCRA), 95–96
Rhine Action Plan (RAP), 135–136
Rhine River, 4, 5, 129–136
  aftermath, 133–136
  authorities' response to incident, 132–133
  fire at Sandoz plant, 129–132
Rio de Janiero Conference (1992). *see* Earth Summit (1992)
Route 66, 91, 100

S
Sakagami, Yuki. *see* Yukijo
Sandoz, 131–135
Schroeder, Karen, 69–70
Scranton, William III, 84, 85
sea otters, 147
seabirds, 148
Seascale, England, 41, 43
Sellafield. *see* Windscale, England
Seveso, Italy, 2, 3, 4, 5, 7, 45–59, 67, 78, 92, 99, 132, 180
  chemicals produced by ICMESA, 45–47
  disposal of dioxin, 55–56
  effects of accident on residents, 47–50
  evacuation, 50–55
  explosion at chemical plant, 47
  long-term effects of accident, 54–59
Seveso Directive, 7, 56, 132
Shashenka, Vladimir, 109
Shimada, Kenichi, 25, 27
Smith, Aileen, 23, 25, 28, 29
Smith, W. Eugene, 23, 25, 28, 29
South Africa, 163–169
South African Foundation for the Conservation of Coastal Birds (SANCCOB), 166, 168
Spill-Sorb, 166–167
spontaneous abortions, 74–75, 108
Stang, Dorothy, 176–177
Steinbeck, John, 91
Stichting Reinwater (Clean Water Foundation), 136
Sting, 177
subsistence living, 137–139, 145, 151–153, 159, 172, 175, 180
Suez Canal, 165
Sujimoto, Eiko, 10
sulfur dioxide, 32, 36, 63, 157
Superfund law, 7, 96–97
Syntex, 95, 99

T
Takeuchi, Tadao, 12
Tanigawa, Gan, 15
tetrachlorodibenzeno-p-dioxin (TCDD). *see* dioxin
thionyl chloride, 62, 63, 69–70
Thornburgh, Richard, 86–87
Three Mile Island, 1, 3, 5, 28, 44, 81–89
  details of incident, 81–84
  evacuations, 86–88
  lawsuits against Met Ed, 88–89
  media coverage, 84–85
  NRC and, 85–88
Times Beach, 3–4, 5, 6, 91–100, 180, 182
  CDC and, 94–95

contamination of area, 92–93
dioxin and, 94–95
EPA and, 95–99
FDA and, 95
hexachlorophene and, 92–93
lawsuits against Bliss, 93–94
Superfund law and, 96–97
TCP and, 94
Trans-Alaska Pipeline Authorization Act (1973), 139
trichlorophenol (TCP), 46, 49, 67–68, 92, 94
Tsurumatsu, Kama, 15–16

U
Ui, Jun, 18
Union Carbide Corporation (UCC), 101–102, 105–107

V
Vietnam War, 46, 92, 94, 156, 161
Voorhees, Aileen and Edwin, 69

W
Wheelwright, Jeff, 147, 148
Wigner energy, 40
Willers, Thomas, 65
Windscale, England, 1, 3, 5, 39–44, 127
World Bank, 177
World Health Organization (WHO), 126
World War I, 9, 62
World War II, 12, 28, 32, 39, 62, 130, 131, 138
World Wildlife Fund, 176, 177, 191

Y
Yukijo, 16–17, 28